Maths Skills *for GCSE*
Science

Carol Tear

OXFORD
UNIVERSITY PRESS

OXFORD
UNIVERSITY PRESS

Great Clarendon Street, Oxford, OX2 6DP, United Kingdom

Oxford University Press is a department of the University of Oxford.
It furthers the University's objective of excellence in research, scholarship,
and education by publishing worldwide. Oxford is a registered trade mark
of Oxford University Press in the UK and in certain other countries

British Library Cataloguing in Publication Data
Data available

978-0-19-843792-5

10 9 8 7 6 5 4 3 2 1

Printed in India by Multivista Global Pvt. Ltd

Acknowledgements
Cover: OUP/grebeshkovmaxim/Shutterstock
p110: Becris/Shutterstock

Artwork by Thomson Digital

Although we have made every effort to trace and contact all
copyright holders before publication this has not been possible in all
cases. If notified, the publisher will rectify any errors or omissions at
the earliest opportunity.

Links to third party websites are provided by Oxford in good faith
and for information only. Oxford disclaims any responsibility for
the materials contained in any third party website referenced in
this work.

2

Contents

How to use this book

This workbook has been written to support the development of key mathematics skills required to achieve success in your GCSE Science course. It has been devised and written by teachers and the practice questions included reflect the AQA, OCR, and Edexcel specifications.

The workbook is structured into chapters with each chapter relating to an area of GCSE Science. Then, each topic covers a mathematical skill or skills that you may need to practise. Each topic offers the following features:

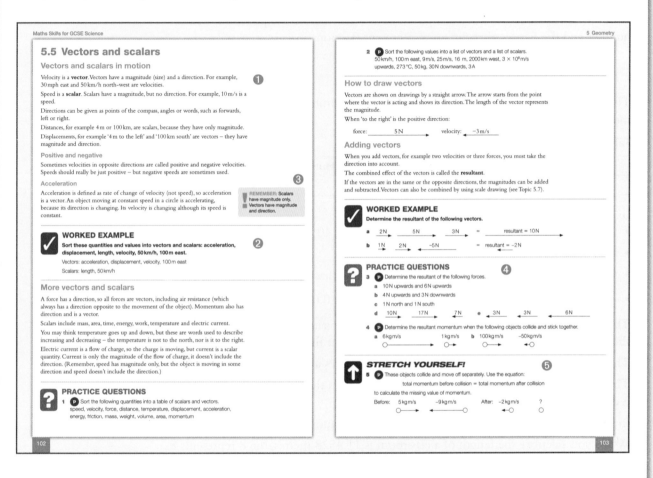

❶ *Opening paragraph* outlines the mathematical skill or skills covered within the topic.

❷ *Worked example* – each topic will have one or two worked examples. The worked examples will be annotated.

❸ *Remember* is a useful box that will offer you tips, hints and other snippets of useful information.

❹ *Practice questions* are ramped in terms of difficulty and all answers are available at www.oxfordsecondary.co.uk

❺ *Stretch yourself* – some of the topics may also contain a few more difficult questions to stretch your mathematical knowledge and understanding.

1 NUMBERS AND UNITS

1.1 Decimal numbers

Numbers: the basics

Whole numbers are called **integers**. For example, 3, 79, 3000. When the answer to a question is not a whole number you can write it as a fraction or a decimal number.

A **fraction**, for example $\frac{1}{2}$ or $\frac{1}{4}$, is a useful way of representing a quantity less than one.

A **decimal number** has a decimal point. Each figure before the point is a whole number and the figures after the point represent fractions.

Place value

The decimal point tells you the 'place value' of the numbers. For example, 4738 means something completely different to 47.38.

Table 1 Place value headings

Tens	Units	Decimal point	Tenths	Hundredths	Thousandths
4	7	•	3	8	○

In this example:

* the 4 is placed in the 'tens' position, so it is worth 40

* the 7 is in the 'units' position, so it is worth 7

* the 3 is in the 'tenths' position, so it is worth 0.3, that is $\frac{3}{10}$

* the 8 is in the 'hundredths' position, so it is worth 0.08, that is $\frac{8}{100}$.

There are no thousandths, so you would not usually write a zero, unless you want to show a certain number of decimal places or significant figures (see Topic 2.1).

The number of decimal places is the number of figures after the decimal point.

The number 47.38 has 2 decimal places, and 47.380 is the same number to 3 decimal places.

In science, you must write your answer to a sensible number of decimal places.

When you calculate the length of a classroom is 11.4528 metres, the 8 represents 8 tenths of a millimetre. The classroom length will vary by more than that when you measure it at different points. Writing it to the nearest centimetre, that is 11.45 metres, is a more sensible answer.

> **REMEMBER:** Only leave your answer as a fraction when the question asks for a fraction. Otherwise calculate the answer as a decimal number.

✓ WORKED EXAMPLES

Different amounts of water were used in an experiment. Here are the results of measurements of water in litres:

1.8 litres, 0.87 litres, 1.452 litres, and 0.628 litres

a List the amounts in order from smallest to largest.

b State the number of decimal places for each measurement.

a Ordering

Start on the left-hand side (LHS) and compare the whole numbers. The smallest are 0.87 and 0.628, as the other numbers start with 1:

1.8 litres, (0.87) litres, 1.452 litres, and (0.628) litres

For 0.87 and 0.628, compare the first figure to the right of the decimal point: 0.6 is less than 0.8, so 0.628 is the smallest. Cross this out and write it as the smallest number in your answer:

1.8 litres, 0.87 litres, 1.452 litres, and ~~0.628 litres~~

0.87 is the next smallest, cross this out and write it as the next number in your answer:

1.8 litres, ~~0.87 litres~~, 1.452 litres, and ~~0.628 litres~~

Compare the remaining whole numbers. Both = 1, so compare the figure to the right of the decimal point:

1.⑧ litres, ~~0.87 litres~~, 1.④52 litres, and ~~0.628 litres~~

4 is smaller than 8, so 1.452 is smaller than 1.8

Answer: 0.628 litres, 0.87 litres, 1.452 litres, 1.8 litres

b The number of decimal places

1.8 litres has 1 decimal place.

0.87 litres has 2 decimal places.

0.628 and 1.452 both have 3 decimal places.

PRACTICE QUESTIONS

1 **B** New antibiotics are being tested. A student calculates the area of clear zones in Petri dishes in which the antibiotics have been used. List these in order from smallest to largest.

   $0.0214\,cm^2$ $0.03\,cm^2$ $0.0218\,cm^2$ $0.034\,cm^2$

2 **B** A student measures the heights of a number of different plants. List these in order from smallest to largest.

   22.003 cm   22.25 cm   12.901 cm   12.03 cm   22 cm

3 **C** A student uses different concentrations of sodium hydroxide in an experiment.

   a List these in order from smallest to largest.

   $20.25\,g/dm^3$ $20.209\,g/dm^3$ $20.009\,g/dm^3$ $20.05\,g/dm^3$

   b List the values above that are written to 2 decimal places.

4 **P** A student measures the current through a component using different batteries. The table shows her results.

Battery	A	B	C	D
Current in A	0.025	0.23	0.087	0.201

Draw a new table with the student's results in order from smallest to largest. Write all of your answers to 3 decimal places.

1.2 Rounding values

Rounding values

You need to round your answer when:

- it is reasonable to give a value 'to the nearest 100' or 'to the nearest whole number'
- the answer to a calculation has many decimal places, for example, $2 \div 3 = 0.666$ recurring, which means there are an infinite number of 6s after the decimal point
- you have been asked to give your answer to 2 decimal places or to 3 decimal places.

Rounding an answer is about deciding which end of the number line the value is closest to.

For example, to decide whether a number is closer to 12 or to 13:

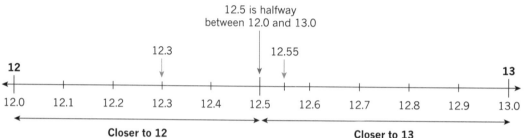

12.5 is halfway between 12.0 and 13.0

12.3   12.55

12   13

12.0   12.1   12.2   12.3   12.4   12.5   12.6   12.7   12.8   12.9   13.0

Closer to 12
Round down numbers where the value is less than halfway to the next number.
12.3 is closer to 12 than 13

Closer to 13
Round up numbers where the value is equal to or more than halfway to, the next number.
12.55 is closer to 13 than 12

To round a number:

- identify the last figure in your rounded answer – it is underlined in the following worked examples
- look at the figure to the right of the underlined answer
- if it is 0, 1, 2, 3, or 4, then do not change the underlined figure
- if it is 5, 6, 7, 8, or 9, then add 1 to the underlined figure.

WORKED EXAMPLES

To the nearest 1̲00:

6̲35 = 600   6̲50 = 700   6̲90 = 700   27̲40 = 2700

27̲50 = 2800   99̲70 = 10000

To 2 decimal places (2 d.p.):

15.33̲33 = 15.33   15.33̲5 = 15.34   15.66̲666 = 15.67

Take care not to round more than once. In calculations with lots of stages, keep the extra figures in your calculator, or write down at least one extra figure in your working until your reach the final answer. This example shows why:

3.454̲545 to 3 d.p. = 3.455   and   3.45̲4545 to 2 d.p. = 3.45

but   3.45̲5 to 2 d.p. = 3.46

PRACTICE QUESTIONS

1 **B** Write the following values to 3 decimal places.

length of a stem = 17.343 43 mm   volume of sugar solution = 1.479 42 ml
mass of potato = 3.4556 g   area of colony = 116.2199 cm²

2 **C** Write the following values to 2 decimal places.

radius of a bromine atom = 0.115 nm   mass of copper = 4.944 g
atomic mass of hydrogen = 1.0079 u   amount of carbon dioxide = 0.995 mol

3 **P** Write the following values to 2 decimal places.

gravitational field strength: g = 9.806 65 N/kg   speed of sound = 0.3385 km/s
voltage of cell = 1.442 V   density of hydrogen = 0.0899 g/cm³

Upper and lower bounds

When you calculate or measure a value, the value may be, for example, to 3 decimal places, or to the nearest whole number. This means there is a range of possible values from an upper bound to a lower bound.

You may need to know the range to work out the answer to a problem.

✓ WORKED EXAMPLE

The length of a tube is 24 cm. This value is to the nearest centimetre (cm), so to the nearest tenth of a cm the actual value could be 23.5 cm, or anywhere between 23.5 cm and 24.5 cm.
It can't be 24.5 because to the nearest cm, that would be 25 cm.
This range can be written 23.5 cm ≤ 24 cm < 24.5 cm.

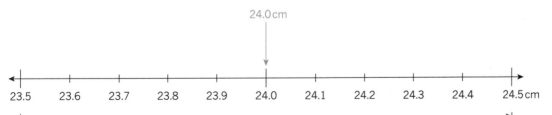

24 cm has a range of values – from 23.5 cm to just below 24.5 cm

> **REMEMBER:**
> < means 'less than'
> ≤ means 'less than or equal to'
> > means 'greater than'
> ≥ means 'greater than or equal to'

PRACTICE QUESTION

4 State the range of possible values for:

a mass = 5 g to the nearest 0.1 g.

b area = 8.5 cm² to nearest 0.01 cm².

1.3 Powers and indices

Writing numbers using powers and indices

Ten squared $= 10 \times 10 = 100$ and can be written as $10^2 = 100$. This is also called 'ten to the power of 2'.

Ten cubed is ten to the power of three and can be written as $10^3 = 1000$.

The little number '2' or '3' is called an index. When we are talking about more than one index number, they are called indices. They show the power the number is raised to, which means the number of times 1 is multiplied or divided by the number. For example:

- ten: $10^1 = 1 \times 10 = 10$
- one hundred: $10^2 = 1 \times 10 \times 10 = 100$

Negative indices

Fractions have negative indices. For example:

- one tenth: $10^{-1} = \dfrac{1}{10} = 0.1$
- one hundredth: $10^{-2} = \dfrac{1}{100} = 0.01$

> **! REMEMBER:** When multiplying numbers you must add the indices.
>
> So $100 \times 1000 = 100\,000$ is the same as
>
> $10^2 \times 10^3 = 10^{2+3} = 10^5$
>
> When dividing numbers you must subtract the indices.
>
> So $100 \div 1000 = \dfrac{1}{10} = 10^{-1}$ is the same as
>
> $10^2 \div 10^3 = 10^{2-3} = 10^{-1}$
>
> But you can only do this when the numbers with the indices are the same.
>
> $10^2 \times 2^3 = 100 \times 8 = 800$
>
> And you can't do this when adding or subtracting.
>
> $10^2 + 10^3 = 100 + 1000 = 1100$
>
> $10^2 - 10^3 = 100 - 1000 = -900$

Indices of 0 or 1

If the index $= 0$, the value is 1. For example, one: $10^0 = 1$ (not multiplied or divided by 10 at all).

You can show that this is true, because one tenth of 10 is 1, which can be written:

$\dfrac{1}{10} \times 10 = 1 \quad$ or $\quad 10^{-1} \times 10^1 = 10^{-1+1} = 10^0$

Any number to the power of 0 is equal to 1, for example, $29^0 = 1$. Don't make the mistake of thinking it is 0.

Notice that if the index is 1, the value is unchanged. For example:

- $10^1 = 10$
- $2^1 = 2$

Writing very large and very small numbers

Indices are useful for extremely large numbers and for extremely small numbers.

- There are approximately 37.2×10^{12} cells in the human body. This is easier to read than $37\,200\,000\,000\,000$ cells.
- There are 6.02×10^{23} molecules of hydrogen in one gram of hydrogen gas. This is $602\,000\,000\,000\,000\,000\,000\,000$ molecules.
- The charge on an electron is $1.6 \times 10^{-19}\,\text{C}$ or $16\,000\,000\,000\,000\,000\,000\,\text{C}$.

WORKED EXAMPLES

1 **Calculate $2^4 \times 2^2$ and write your answer a with and b without using indices.**

   a $2^4 \times 2^2 = 2^{4+2} = 2^6$

   b $2 \times 2 \times 2 \times 2 \times 2 \times 2 = 64$

2 **Calculate $10^3 \div 10^{-6}$ and write your answer without using indices.**

   $$10^3 \div 10^{-6} = 10^{3-(-6)}$$
   $$= 10^{3+6}$$
   $$= 10^9$$
   $$= 1\,000\,000\,000$$

PRACTICE QUESTIONS

1 Calculate the following values. Give your answers using indices.

   a $10^8 \times 10^3$

   b $10^7 \times 10^2 \times 10^3$

2 Calculate the following values. Give your answers with and without using indices.

   a $10^5 \div 10^4$

   b $10^3 \div 10^6$

   c $10^2 \div 10^{-2}$

   d $10^2 \div 10^{-4}$

   e $2^8 \div 2^5$

   f $100^2 \div 10^2$

3 Calculate the following values – read the questions very carefully!

   a $10^3 + 10^3$

   b $10^2 - 10^{-2}$

   c $2^3 \times 10^2$

1.4 Standard form

Standard form

Standard form is where the whole part of the number is a value from 1 to 9.

The following numbers are written in standard form.

- There are 6.02×10^{23} molecules of hydrogen in one gram of hydrogen gas.
- The charge on an electron is 1.6×10^{-19} C.

The following numbers are not written in standard form.

- The density of aluminium is $2700\,\text{kg/m}^3$.
- There are approximately 37.2×10^{12} cells in the human body.

In standard form, the density of aluminium $= 2.7 \times 10^3\,\text{kg/m}^3$.

2.7 is 1000 times smaller than 2700, so 2.7 must be multiplied by 10^3 to keep the value the same:

$$2700 = 2.7 \times 10^3$$

The number of cells in the human body is approximately 3.72×10^{13}.

3.72 is ten times smaller than 37.2, so 10^{12} must be multiplied by ten to keep the number the same:

$$10^{12} \times 10 = 10^{13}$$
$$37.2 \times 10^{12} = 3.72 \times 10^{13}$$

WORKED EXAMPLES

Write the following in standard form.

a   15 000 000 microorganisms.

15 000 000 in standard form will have one figure before the decimal point so it will be $1.5 \times 10^{\text{something}}$.

Count the number of figures after this decimal point:

number of figures = 7. The decimal point has moved 7 places to the left.

$= 1.5 \times 10^7$ microorganisms

b   The mass of a neutron: 0.01675×10^{-25} kg.

In standard form, this will be $1.675 \times 10^{-\text{something}}$ kg.

Count the number of figures the decimal point has moved:

number of figures = 2. The decimal point has moved 2 places to the right.

$= 1.675 \times 10^{-2} \times 10^{-25} = 1.675 \times 10^{-2 + (-25)}\,\text{kg} = 1.675 \times 10^{-27}\,\text{kg}$

c **The diameter of a copper atom: 0.000 228 millionths of a metre.**

1 millionth $= 10^{-6}$

0.000 228 millionths of a metre $= 0.000\,228 \times 10^{-6}\,\text{m}$. In standard form, this will be $2.28 \times 10^{-\text{something}}\,\text{m}$.

Count the number of figures the decimal point has moved:

$0.0002.28 \;= 2.28 \times 10^{-4} \times 10^{-6} = 2.28 \times 10^{-4\,+\,(-6)} = 2.28 \times 10^{-10}\,\text{m}$

1234

PRACTICE QUESTIONS

1 **B** Change the following values to standard form.

a energy in 100 g of butter: 3060 kJ

b mass of a blue whale: 140 000 kg

c diameter of a human hair: 0.000 18 m

d length of an *E. coli* bacterium: 0.000 004 m

2 **C** Change the following values to standard form.

a boiling point of sodium chloride: 1413 °C

b mass of a proton: $1673 \times 10^{-30}\,\text{kg}$

c largest nanoparticles: $0.0\,001 \times 10^{-3}\,\text{m}$

d number of atoms in 1 mol of water: 1806×10^{21}

3 **P** Change the following values to standard form.

a density of gold: 19 000 kg/m³

b speed of light: 300 000 000 m/s

c atmospheric pressure: 101 000 Pa

d half-life of carbon-22: 0.0062 s

STRETCH YOURSELF!

4 **B** Use the formula:

$$\text{actual size} = \frac{\text{image size}}{\text{magnification}}$$

to calculate the actual length of an image 150 mm long when viewed with a magnification of × 500. Write your answer in standard form **a** in mm **b** in m.

5 **C** The radius of a nucleus is about $\dfrac{1}{10\,000}$ times the radius of the atom. The radius of the atom is about $1 \times 10^{-10}\,\text{m}$. Calculate the radius of the nucleus in standard form.

6 **P** Use the formula:

$$\text{length} \times \text{width} \times \text{height}$$

to calculate the volume of a block 2 cm × 6 cm × 3 cm. Write your answer:
a in cm³   **b** in standard form in m³.

1.5 Prefixes

Prefixes for large numbers

Prefixes are used with Système International (SI) units (see Topic 1.6) when the value is very large or very small. They can be used instead of writing the number in standard form (see Topic 1.4).

Table 1 shows prefixes for large numbers. For example:

- A kilowatt (1 kW) is a thousand watts, that is 1000 W or 10^3 W.

- A megawatt (1 MW) is a million watts, that is 1 000 000 W or 10^6 W.

- A gigawatt (1 GW) is a billion watts, that is 1 000 000 000 W or 10^9 W.

Gansu Wind Farm in China has an output of 6.8×10^9 W. This can be written as 6800 MW or 6.8 GW.

Table 1 Prefixes for large numbers

Prefix	Symbol	Value	Prefix	Symbol	Value
kilo	k	$10^3 = 1000$	giga	G	$10^9 = 1\,000\,000\,000$
mega	M	$10^6 = 1\,000\,000$	tera	T	$10^{12} = 1\,000\,000\,000\,000$

! **REMEMBER:** Except for k, the symbols are all upper case. The factors increase by one thousand. So 1 GW = 1000 MW.

WORKED EXAMPLES

A radio station broadcasts radio waves with a frequency of 103.7 MHz. Write this in a Hz and b standard form.

a 103.7 MHz = 103.7×10^6 Hz

$= 103.7 \times 1\,000\,000$ Hz

$= 103\,700\,000$ Hz

b 103.7×10^6 Hz = $1.037 \times 10^2 \times 10^6$ Hz

$= 1.037 \times 10^{2\,+\,6}$ Hz

$= 1.037 \times 10^8$ Hz

! **REMEMBER:** If you are not sure whether to divide or multiply, use the statement below to help you.

Your answer will be bigger if the unit is smaller – there will be 1000 times more grams than kilograms, and 1000 times fewer gigawatts than kilowatts.

PRACTICE QUESTIONS

1 **B** A burger contains 4 500 000 J of energy. Write this in:

   a kilojoules   b megajoules.

2 **C** A car emits about 1 530 000 g of carbon dioxide in a year. Write this in:

   a kg   b standard form.

3 **P** The latent heat of vaporisation of water is 2 260 000 J/kg. Write this in:

   a J/g   b kJ/kg   c MJ/kg.

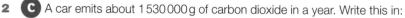

Prefixes for small numbers

As well as prefixes for large numbers there are prefixes for small numbers that are fractions of a unit.

Table 2 shows prefixes for small numbers. For example:

- There are a billion nanometres in a metre, that is $1\,000\,000\,000\,\text{nm} = 1\,\text{m}$.
- There are a million micrometres in a metre, that is $1\,000\,000\,\mu\text{m} = 1\,\text{m}$.

A red blood cell has a diameter of about $7 \times 10^{-6}\,\text{m}$. This can be written as $7\,\mu\text{m}$.

Table 2 Prefixes for small numbers

Prefix	Symbol	Value	Prefix	Symbol	Value
deci	d	$10^{-1} = \dfrac{1}{10}$	micro	µ	$10^{-6} = \dfrac{1}{1\,000\,000}$
centi	c	$10^{-2} = \dfrac{1}{100}$	nano	n	$10^{-9} = \dfrac{1}{1\,000\,000\,000}$
milli	m	$10^{-3} = \dfrac{1}{1000}$			

> **REMEMBER:** The symbols are all lower case. Micro has the Greek symbol µ pronounced 'mew'. Except for deci and centi, the factors decrease by one thousand. So $1\,\mu\text{g} = \dfrac{1}{1000}\,\text{g}$.

WORKED EXAMPLE

Change $0.05\,\text{m}^3$ to mm^3.

There are 1000 or $10^3\,\text{mm}$ in $1\,\text{m}$.

A metre cube will have length, breadth, and height $1\,\text{m} = 10^3\,\text{mm}$.

Volume of $1\,\text{m}^3 = 10^3\,\text{mm} \times 10^3\,\text{mm} \times 10^3\,\text{mm} = 10^{3+3+3}\,\text{mm}^3 = 10^9\,\text{mm}^3$.

$0.05\,\text{m}^3 = 0.05 \times 10^9\,\text{mm}^3 = 5 \times 10^7\,\text{mm}^3$

PRACTICE QUESTIONS

4 **B** HIV is a virus with a diameter of between $9.0 \times 10^{-8}\,\text{m}$ and $1.20 \times 10^{-7}\,\text{m}$. Write this range in nanometres.

5 **C** Write the following volumes in cm^3.

    a $1\,\text{m}^3$ **b** $1\,\text{dm}^3$ **c** $100\,\text{mm}^3$

6 **P** Write the following measurements using suitable prefixes.

    a a microwave wavelength $= 0.009\,\text{m}$

    b a wavelength of infrared $= 1 \times 10^{-5}\,\text{m}$

    c a wavelength of blue light $= 4.7 \times 10^{-7}\,\text{m}$

1.6 Units

SI units

Most of the numbers you use in science are amounts of something and have a unit. In science most units are Système International (SI) units. Examples are the kilogram, metre, and second. It is very important that your answers include the units, otherwise it is not clear whether the number means, for example, grams, kilometres, or hours.

Table 1 shows the units used in GCSE Science. The name of the unit is always lower case. The symbol must be as shown in the table – some are lower case and some are upper case.

Table 1 Units used in GCSE Science

Base SI units

Quantity	Unit	Symbol	Quantity	Unit	Symbol
length	metre	m	electric current	ampere	A
mass	kilogram	kg	temperature	kelvin	K
time	second	s	the amount of substance	mole	mol

Derived SI units and other units (non SI in purple)

Quantity	Unit	Symbol	Quantity	Unit	Symbol
acceleration	metre per squared second	m/s^2	gravitational field strength	newton per kilogram	N/kg
activity	becquerel	Bq	power	watt	W
area	squared metre	m^2	pressure	pascal	Pa
concentration	gram per cubic decimetre, mole per cubic decimetre	g/dm^3, mol/dm^3	rate of reaction	gram per second, cubic centimetre per second, mole per second	g/s, cm^3/s mol/s
density	kilogram per cubic metre	kg/m^3	specific heat capacity	joule per kilogram per degree Celsius	J/kg °C
electric charge	coulomb	C	specific latent heat	joule per kilogram	J/kg
electric potential difference	volt	V	speed	metre per second, kilometre per hour	m/s km/h
energy	joule	J	temperature	degree Celsius	°C
force	newton	N	volume	cubic metre, litre, cubic decimetre	m^3, l or L, dm^3
frequency	hertz	Hz	gravitational field strength	newton per kilogram	N/kg

Converting units

In Table 1, you can see that some quantities can be measured in more than one unit. For example, speed is measured in metres per second and in kilometres per hour. A car may be travelling at 100 km/h and you might need to work out what this is in metres per second.

Another conversion you may need is volume, from cubic decimetres or cubic metres to litres.

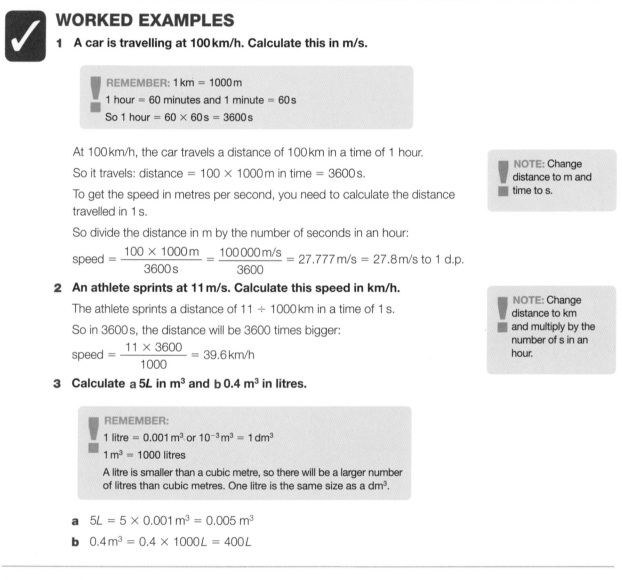

WORKED EXAMPLES

1 **A car is travelling at 100 km/h. Calculate this in m/s.**

> **REMEMBER:** 1 km = 1000 m
> 1 hour = 60 minutes and 1 minute = 60 s
> So 1 hour = 60 × 60 s = 3600 s

At 100 km/h, the car travels a distance of 100 km in a time of 1 hour.

So it travels: distance = 100 × 1000 m in time = 3600 s.

To get the speed in metres per second, you need to calculate the distance travelled in 1 s.

So divide the distance in m by the number of seconds in an hour:

$$\text{speed} = \frac{100 \times 1000\,\text{m}}{3600\,\text{s}} = \frac{100\,000\,\text{m/s}}{3600} = 27.777\,\text{m/s} = 27.8\,\text{m/s to 1 d.p.}$$

> **NOTE:** Change distance to m and time to s.

2 **An athlete sprints at 11 m/s. Calculate this speed in km/h.**

The athlete sprints a distance of 11 ÷ 1000 km in a time of 1 s.

So in 3600 s, the distance will be 3600 times bigger:

$$\text{speed} = \frac{11 \times 3600}{1000} = 39.6\,\text{km/h}$$

> **NOTE:** Change distance to km and multiply by the number of s in an hour.

3 **Calculate a 5L in m³ and b 0.4 m³ in litres.**

> **REMEMBER:**
> 1 litre = 0.001 m³ or 10^{-3} m³ = 1 dm³
> 1 m³ = 1000 litres
> A litre is smaller than a cubic metre, so there will be a larger number of litres than cubic metres. One litre is the same size as a dm³.

a $5L = 5 \times 0.001\,\text{m}^3 = 0.005\,\text{m}^3$

b $0.4\,\text{m}^3 = 0.4 \times 1000L = 400L$

PRACTICE QUESTIONS

1 Calculate the following speeds in m/s.

   a a jet airliner speed of 880 km/h   **b** a cyclist cycles at 16 km/h

2 Calculate the following speeds in km/h.

   a speed of sound 330 m/s   **b** a horse gallops at 12.5 m/s

3 Calculate the following volumes.

   a 300L in m³   **b** 500 ml in dm³   **c** 0.01 m³ in L   **d** 15 dm³ in L

1.7 Fractions and percentages

Fractions

A fraction is a way of representing a part of something. When you divide something to get four equal parts, each part is a quarter. This can be written as:

$\frac{1}{4}$ ◄— numerator
◄— denominator   This means $1 \div 4$, which is the same as one quarter.

Which is larger: $\frac{7}{25}$ or $\frac{3}{10}$? It is not easy to see without a calculation.

You can only compare the numerators (the 7 and the 3 on the top) when the denominators (the 25 and the 10 on the bottom) are the same. To change the denominator without changing the value of the fraction, you must multiply the top and the bottom by the same number. Both these fractions can be changed to a number of hundredths:

$$\frac{7 \times 4}{35 \times 4} = \frac{28}{100} \quad \text{and} \quad \frac{3 \times 10}{10 \times 10} = \frac{30}{100}$$

30 is larger than 28, so $\frac{3}{10}$ is larger than $\frac{7}{25}$.

Writing fractions as percentages

In the example above, it was easier to compare the two fractions once they had been converted to hundredths. This is why scientists use percentages. You can think of 'per cent' as meaning 'hundredths'.

In the example above: $\frac{7}{25} = \frac{28}{100} = 28\%$

If you have not already learnt the basic fractions as percentages, it is worth doing so:

$75\% = \frac{3}{4}$ $50\% = \frac{1}{2}$ $25\% = \frac{1}{4}$ $20\% = \frac{1}{5}$ $10\% = \frac{1}{10}$ $1\% = \frac{1}{100}$

A percentage does not have to be a whole number. For example: $\frac{1}{3} = 33\frac{1}{3}\%$ or 33.33%.

To work out a percentage, you must identify or calculate the total number:

$$\text{percentage} = \frac{\text{number you want as a percentage}}{\text{total number}} \times 100\%$$

REMEMBER:
Only leave your answer as a fraction if you have been asked to give your answer as a fraction, or you may lose marks for not completing the calculation.

✓ WORKED EXAMPLES

1   The total biomass of plant life in a pond is 2000 g/m². The total biomass of the primary consumers is 700 g/m². Calculate the percentage of biomass that is passed to the primary consumers.

$$\text{Percentage} = \frac{\text{total biomass of the primary consumers}}{\text{total biomass of plant life in a pond}} \times 100\%$$

$$\text{Percentage} = \frac{700}{2000} \times 100\% = 35\%$$

2 Calculate the percentage by mass of carbon in carbon dioxide. (Relative mass of carbon = 12, of oxygen = 16)

$$\text{Percentage by mass} = \frac{\text{relative mass of carbon}}{\text{total relative formula mass of carbon dioxide}} \times 100\%$$

Formula of carbon dioxide = CO_2 = 1 carbon atom + 2 oxygen atoms

Total relative formula mass of 1 molecule = $12 + (2 \times 16) = 44$

$$\text{Percentage} = \frac{12}{44} \times 100\%$$

$$= 27.27\% = 27\% \text{ (2 s.f.)}$$

> **NOTE:** See Topic 2.1 for more information on significant figures (s.f.).

3 A thermal power station uses 11 600 kWh of energy from fuel to generate electricity. A total of 4500 kWh of energy is output as electricity. Calculate the percentage of energy 'wasted' (dissipated in heating the surroundings).

$$\text{Percentage energy wasted} = \frac{\text{energy wasted}}{\text{total energy input from fuel}} \times 100\%$$

You must calculate the energy wasted using the value for useful energy output:

$$\text{Percentage energy wasted} = \frac{\text{total energy input from fuel} - \text{energy output as electricity}}{\text{total energy input from fuel}} = \times 100\%$$

Energy wasted = (11 600 − 4500) kWh

$$\text{Percentage energy wasted} = \frac{(11600 - 4500)}{11600}$$

$$= 61.2\% = 61\% \text{ (2 s.f.)}$$

PRACTICE QUESTIONS

1 **B** The table below shows some data about energy absorbed by a tree in a year and how some of it is transferred.

Energy absorbed by the tree in a year	3 600 000 kJ/m²
Energy transferred to primary consumers	2240 kJ/m²
Energy transferred to secondary consumers	480 kJ/m²

Calculate the percentage of energy absorbed by the tree that is transferred to **a** primary consumers **b** secondary consumers.

2 **P** Calculate the efficiency of a motor that does 8400 J of work to lift a load. The electrical energy supplied is 11 200 J.

3 **P** A student measures the potential difference across a component to be 2.5 ± 0.2 V. (This means there is an uncertainty in the value and it could be anywhere between 2.5 − 0.2 = 2.3 V and 2.5 + 0.2 = 2.7 V.) Calculate the percentage uncertainty in the potential difference. Write your answer as 2.5 V ± .....%.

4 **C** Calculate the percentage by mass of **a** copper in copper sulfide CuS

   b sulfur in copper sulfate, $CuSO_4$. (Relative mass of copper = 63, of sulfur = 32, and of oxygen = 16.)

5 **C** Use the equation:

relative atomic mass = sum of (fraction of isotope × relative atomic mass of isotope) for all isotopes

to calculate the relative atomic mass of boron (boron-10 has 20% abundance boron-11 has 80% abundance) to 3 s.f.

1.8 Percentage calculations

Writing changes as percentages

When you work out an increase or a decrease as a percentage change, you must identify, or calculate, the total original amount:

$$\text{percentage increase} = \frac{\text{increase}}{\text{original total amount}} \times 100\%$$

$$\text{percentage decrease} = \frac{\text{decrease}}{\text{original total amount}} \times 100\%$$

> **! REMEMBER:** When you calculate a percentage change, use the total before the increase or decrease, not the final total.

WORKED EXAMPLES

1 In 2001, 3.2 million tonnes of UK household waste was collected for recycling. By 2016, this had increased to 10 million tonnes. Calculate the percentage increase in household waste recycling.

Increase = tonnes in 2016 − tonnes in 2001

Increase = 10 million tonnes − 3.2 million tonnes = 6.8 million tonnes

Original total = 3.2 million tonnes

Percentage increase $= \dfrac{6.8}{3.2} \times 100\%$

$\qquad\qquad = 212.5\% = 210\%$ (2 s.f.)

2 The electricity generated from coal in the UK decreased from 76 000 GWh in 2015 to 31 000 GWh in 2016. Calculate the percentage decrease in electricity generation from coal.

Decrease = GWh in 2015 − GWh in 2016

Decrease = 76 000 GWh − 31 000 GWh = 45 000 GWh

Original total = 76 000 GWh

Percentage decrease $= \dfrac{45\,000}{76\,000} \times 100\%$

$\qquad\qquad = 59.2\% = 59\%$ (2 s.f.)

PRACTICE QUESTIONS

1 (C) UK air pollution statistics show that sulfur dioxide emissions have fallen from 0.77 million tonnes in 2006 to 0.18 million tonnes in 2016. Calculate the percentage decrease in sulfur dioxide emissions.

2 (P) At 50 km/h the stopping distance for a car is 35 m and at 60 km/h it is 45 m. Calculate the percentage increase in stopping distance.

3 (B) In the UK in 1971, 22% of patients with bowel cancer survived for 10 years or more. In 2011, this had increased to 57%. Calculate the percentage increase in survival rate.

Calculating the percentage of a value

There are different methods for working out a percentage of a value. Choose the one that works for you.

WORKED EXAMPLE

Calculate 65% of 76.

Method 1

This is a useful method to use when you don't have a calculator, and for simple values such as 50% or 20%.

65% = 50% + 10% + 5%

$50\% = \frac{1}{2}$ of 76 = 38

$10\% = \frac{1}{10}$ of 76 = 7.6

$5\% = \frac{1}{2}$ of 10% (or $\frac{1}{10}$ of 50%) = 3.8

50% + 10% + 5% = 38 + 7.6 + 3.8 = 49.4 = 49 (2 s.f.)

Method 2

$65\% = \frac{65}{100}$

$\frac{65}{100} \times 76 = 49.4 = 49$ (2 s.f.)

PRACTICE QUESTIONS

4　**B**　Salmon is 20% protein and 11% fat by mass. Calculate the mass of **a** protein and **b** fat in a 180 g portion of salmon.

5　**C**　Common brass is made of copper and zinc. It is 63% copper. Calculate the mass of **a** copper and **b** zinc in a 540 g sample.

6　**P**　A solar panel has an efficiency of 15%. It receives 1200 kWh of energy from the Sun. Calculate the useful energy transferred electrically.

7　**P**　The Sankey diagram below shows the energy input and output for an LED light. Calculate the energy transferred by **a** light and **b** the wasted energy transferred directly to the thermal store of the surroundings.

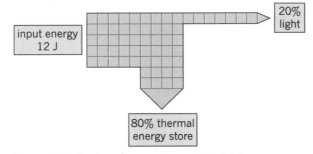

Figure 1 A Sankey diagram for an LED light

8　**C**　The maximum theoretical yield from a reaction is 4.8 g, but when a student does the experiment their percentage yield is 82%. Calculate the actual yield.

1.9 More percentages

Calculating increases and decreases

When you work out an increase or a decrease as a percentage change, you must identify, or calculate, the total original amount:

$$\% \text{ increase} = \frac{\text{increase}}{\text{total original amount}} \times 100$$

$$\% \text{ decrease} = \frac{\text{decrease}}{\text{total original amount}} \times 100$$

WORKED EXAMPLE

150 g of a solid is heated and a gas is given off. The remaining solid has a mass of 84 g. Calculate the percentage decrease in mass.

Decrease in mass = 150 − 84 = 66 g

Original mass = 150 g

$\% \text{ decrease} = \dfrac{66}{150} \times 100 = 44\%$

Calculating the total amount

You are given a percentage and asked to find the original value.

For example, 50% of a number = 40.

You must calculate the original number that was 100%.

$\dfrac{50}{100} \times$ the number = 40, so the original number $= \dfrac{100}{50} \times 40 = 80$

In this example, 50% is one half. So if one half of the number is 40, the original number was double 40 = 80.

WORKED EXAMPLE

A family estimates they use 23% of their electricity for heating and this is 5000 kWh per year. Calculate their total electricity use in a year.

Method 1

$23\% = \dfrac{23}{100}$ of the total = 5000 kWh

1% = 5000 ÷ 23 = 217 kWh

100% = 217 × 100 = 21 700 kWh

Method 2

$\dfrac{23}{100} \times$ total = 5000 kWh

total $= \dfrac{5000 \times 100}{23} = 21\,700$ kWh

PRACTICE QUESTIONS

1   **B** Some students count the number of different species in a pond. There are 19 different species. A year later they repeat the count and there are 23 species. Calculate the percentage increase in the number of different species.

2   **C** A student compares the rate of reaction for a chemical reaction at two different temperatures. At the low temperature, the rate is $15\,cm^3$ gas produced in 1 minute. At the higher temperature, the rate is $17\,cm^3$ gas produced in 1 minute. Calculate the percentage increase in the reaction rate.

3   **P** Some energy suppliers predict that having a smart meter for gas and electricity will lead to a saving of about 450 kWh per year. The total energy use for a household is about 11 000 kWh per year. Calculate the predicted percentage decrease.

4   **B** In 2018, there were 35 200 breeding pairs of puffins on the Farne Islands – 88% of the number in 2000. Calculate the number of pairs in 2000.

5   **C** In an industrial process, the percentage yield from a reaction is 82%. The actual yield is 460 kg. Calculate the maximum theoretical yield.

6   **P** An offshore wind turbine delivers 5.2 MW of power. Its efficiency is 48%. Calculate the power of the wind striking the blades.

Other types of percentage question

1   Find the new value after a percentage increase or a decrease.

2   Find the value before a percentage increase or a decrease.

WORKED EXAMPLES

1   **a**   450 increases by 25%: $100\% + 25\% = 125\%$ of $450 = \dfrac{125}{100} \times 450$

$$= 560 \ (2 \text{ s.f.})$$

   b   450 decreases by 25%: $100\% - 25\% = 75\%$ of $450 = \dfrac{75}{100} \times 450$

$$= 337.5$$

$$= 340 \ (2 \text{ s.f.})$$

2   **a**   After 25% is added, you have 450. Calculate the original number.

$450 = 100\% + 25\% = 125\%$

100% of $450 = \dfrac{100}{125} \times 450 = 360$

   b   After 25% is subtracted, you have 450. Calculate the original number.

$450 = 100\% - 25\% = 75\%$

100% of $450 = \dfrac{100}{75} \times 450 = 600$

STRETCH YOURSELF!

7   **P** The most powerful motor, built to power ships, is very efficient. It delivers 36.5 MW of output power and the power wasted is only about 2% of the input power.

Calculate the input power of the motor to 3 s.f.

1.10 Ratios

Writing ratios

If you want to compare two quantities – or two numbers of anything – one way to do this is to write it as a ratio. $1:2$ (one to two) is a ratio and $6:1$ (six to one) is a ratio.

For example, two dogs breed and have 3 black puppies and 1 yellow puppy. The ratio of black to yellow puppies is $3:1$.

Over the years, the dogs have 24 black puppies and 8 yellow puppies. This ratio is $24:8$.

Both 24 and 8 can be divided by 8 to give:

$$\text{ratio} = \frac{24}{8}:\frac{8}{8} = 3:1$$

Write ratios in the form: 'number':1 or 1:'number'

If the ratio was $26:8$, you can still divide both numbers by 8 to get the ratio $3.25:1$.

> **REMEMBER:** Be careful to write ratios the correct way round.
> - The ratio of black to yellow puppies is $3:1$
> - The ratio of yellow to black puppies $= 1:3$

WORKED EXAMPLES

1 **Write the ratio of hydrogen to oxygen atoms in a molecule of water.**

   H_2O = 2 hydrogen and 1 oxygen atoms

   Ratio $= 2:1$

2 **Write the ratio of oxygen atoms to carbon atoms and hydrogen atoms in a glucose molecule ($C_6H_{12}O_6$).**

   $C_6H_{12}O_6$ = 12 hydrogen, 6 carbon, and 6 oxygen atoms.

   The ratio asked for is in a different order to the atoms in the formula: ratio
   $O:C:H = 6:6:12$

   All three numbers can be divided by 6: ratio $= 1:1:2$

3 **In a survey, the number of left-handed people was 43 and right-handed was 397. Write the ratio of left-handed to right-handed people.**

   left $= 43$ right $= 397$ ratio left:right $= 43:397$

   Divide 43 and 397 by 43:

   $$\text{ratio} = \frac{43}{43}:\frac{397}{43} = 1:9.2$$

PRACTICE QUESTIONS

1 **C** Write the ratio of copper to sulfur and oxygen atoms in a molecule of copper sulfate ($CuSO_4$).

2 **B** Two cats breed several times. Overall, 15 kittens have short hair and 5 have long hair. Write the ratio of kittens with long hair to kittens with short hair.

3 **P** Write the turns ratio (turns in primary coil:turns in secondary coil) for a transformer with 5000 turns on the primary coil and 200 turns on the secondary coil.

4 **B** Write the surface to volume ratio of a cell with a surface area of 201 μm^2 and a volume of 268 μm^3.

Using ratios in calculations

Ratios can be useful in calculations when something has a fixed ratio, such as the number of atoms in a molecule.

WORKED EXAMPLES

1 **A sample of calcium chloride ($CaCl_2$) contains 17 913 atoms. Calculate the number of calcium and the number of chlorine atoms.**

The ratio of calcium to chlorine atoms is $1:2$. To change this ratio to fractions, add the two numbers together: $1 + 2 = 3$. For every 3 atoms, 1 is calcium and 2 are chlorine. This is the same as saying $\frac{1}{3}$ of the atoms are calcium and $\frac{2}{3}$ are chlorine.

Calcium atoms $= \frac{1}{3} \times 17\,913 = 5971$ Chlorine atoms $= \frac{2}{3} \times 17\,913 = 11\,942$

2 **A sample of phosphine gas contains 77.5 g of phosphorus and 7.5 g of hydrogen. Calculate the formula of phosphine.**

From the periodic table (see page 119), relative formula masses of $P = 31\,g$ and $H = 1\,g$.

The number of relative formula masses in the sample are:

for $P = \dfrac{77.5g}{31g} = 2.5$ for $H = \dfrac{7.5g}{1g} = 7.5$

So the ratio of phosphorus atoms to hydrogen atoms, $P:H = 2.5:7.5$

Divide both by 2.5: ratio $P:H = 1:3$

The formula of phosphine $= PH_3$

PRACTICE QUESTIONS

5 **B** The ratio of rabbits to foxes is $5:2$. There are 680 foxes. Calculate the number of rabbits.

6 **P** The gear ratio is:

the number of turns of **driven** gear : the number of turns of **driver** gear

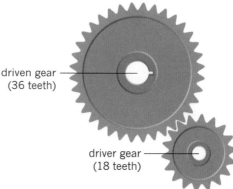

driven gear (36 teeth)

driver gear (18 teeth)

a Write the gear ratio for the gears in the diagram.

b The **driver** gear makes 6 revolutions. Calculate the number of revolutions of the **driven** gear.

7 **C** A sample of toluene contains 336 g of carbon and 32 g of hydrogen. Calculate the formula of toluene.

STRETCH YOURSELF!

8 **C** Water is formed in this reaction: $2H_2 + O_2 = 2H_2O$. A student burns 16 dm³ of hydrogen gas at 0 °C and 1 atmosphere pressure.

a Calculate the volume of oxygen used.

b Calculate the mass of water formed. (Hint: consider the volume of water vapour (gas) formed.)

1.11 Direct proportion

Directly proportional relationships

Sometimes the ratio between things changes. For example, the ratio of rabbits to foxes changes every year.

Sometimes the ratio between two things is constant and does not change. For example, the hydrogen atoms in a molecule of water are always in the ratio:

water molecules 1 : hydrogen atoms 2.

When the ratio is constant, if you double the number of water molecules, you will double the number of hydrogen atoms.

This means the number of hydrogen atoms in water **is directly proportional to** the number of water molecules:

number of hydrogen atoms in water ∝ number of water molecules

The symbol ∝ means 'is directly proportional to'.

When you double the force stretching a spring, you will double the extension of the spring – as long as you don't overstretch the spring with too much force.

This means the extension of the spring **is directly proportional to** the stretching force – as long as the spring is not overstretched. (This is known as Hooke's Law and the limit where it stops applying because the spring is overstretched is called the Hooke's Law limit.)

$$F \propto x$$

Where F is the force on a spring and x is the extension of the spring.

You can also write this as an equation:

$$F = kx$$

Where k has a constant value for this spring and is called the spring constant.

Rearranging this equation (see Topic 3.4):

$$\frac{F}{x} = k$$

WORKED EXAMPLE

Tables 1, 2, and 3 show the current through three electrical components when different potential differences are applied across them. For each component, use the table to judge whether the current is proportional to potential difference (pd) and describe what these results tell you about the resistance of the components.

Table 1

Resistance 1	
pd in V	current in mA
0	0
2.0	8
4.0	16
8.0	32

Table 2

Resistance 2		
pd in V	current in mA	V/I in Ω
0	0	–
3.0	6	500
5.0	10	500
7.0	14	500

Table 3

Filament lamp		
pd in V	current in mA	V/I in Ω
0	0	–
3.0	12	250
5.0	18	278
7.0	22	318

Resistance 1

Table 1 shows that each time the pd doubles, the current doubles. This means that:

$$pd \propto current$$

If pd = V and current = I, then:

$$V \propto I \quad \text{and} \quad \frac{V}{I} = constant$$

The equation for resistance R is:

$$R = \frac{V}{I}$$

So resistance R is constant and using a pair of values:

$$R = \frac{4.0V}{16mA} = \frac{4.0}{0.16} = 250\,\Omega$$

Resistance 2

The relationship between pd and current is not so obvious because there are no values in Table 2 that double.

If $V \propto I$, then $\frac{V}{I}$ = constant.

The extra column added to Table 2 shows that $\frac{V}{I}$ = constant = 500.

So pd $\propto$ current, and each of these pairs of values gives the resistance $R = 500\,\Omega$.

Filament lamp

The same treatment shows that V is not directly proportional to current because as the pd and current change, the resistance changes too.

The resistance is still given by $R = \frac{V}{I}$, but it is not constant and increases as both pd and current increase.

There is a correlation between pd and current. Current increases when pd increases. At larger values of pd, the increase in current is less for the same increase in pd (Table 3).

> **NOTE:** For more about direct and inverse proportion, see Topics 4.3 and 4.4. For more about describing correlations, see Topic 2.11.

PRACTICE QUESTIONS

1 **P** Use your knowledge of physics equations to explain whether the following quantities are directly proportional.

   a   Force and acceleration of a spaceship in deep space

   b   The mass and volume of lead

   c   The volume and length of the side of a cube

   d   The kinetic energy of an object and its speed

1.12 Direct and inverse proportion

Direct proportion

The proportion of left-handed people in the world is approximately 10%.

Out of 100 people, about 10 will be left-handed. Out of 200 people, about 20 will be left-handed. Out of 400 people, about 40 will be left-handed. This is called **direct proportion** as both quantities are increasing. If you double one, the other also doubles.

If you decrease the volume of water in a beaker you will decrease the mass of water. This is another example of direct proportion:

$$\text{mass} \propto \text{volume}$$

PRACTICE QUESTIONS

1   **C** Two containers have the same volume of reactants. In one container, a 0.2 molar solution of hydrochloric acid is mixed with a 0.2 molar solution of sodium hydroxide. In the second container, a 0.4 molar solution of hydrochloric acid is mixed with a 0.4 molar solution of sodium hydroxide. Explain how you expect the concentration to affect the rate of reaction and whether the rate of reaction is proportional to the concentration.

2   **P** Table 4 shows the mass of solid aluminium that is melted by different amounts of energy.

Table 4

Aluminium	
Energy in kJ	**Mass in kg**
320	0.8
600	1.5
880	2.2

   a   Describe how the mass that is melted changes with the amount of energy transferred.

   b   Calculate the specific latent heat. (Choose an equation from the lists of equations on pages 116–118.)

3   **P** When a spring is stretched by a force of 4.8 N, it extends 25 mm. Calculate the extension when it is stretched by a force of **a** 2.4 N **b** 9.6 N. (Choose an equation from the lists of equations on pages 116–118.)

4   **C** Explain whether the following quantities are directly proportional.

   a   The mass of 2 moles of a compound and its relative formula mass

   b   The number of moles of a gas and its volume

   c   The surface area to volume ratio for a cube

Inverse proportion

When the distance from a light source is increased, the light intensity decreases. This is an inverse relationship.

For example, the rate of photosynthesis depends on light intensity. You may be investigating this by increasing the distance between a lamp and a plant so that you decrease the light intensity.

As the lamp moves further away from the plant, the light is spread over a larger area. As the diagram shows, when you double the distance from the lamp, the light is spread over an area 4 times the size. So a square of the same area only receives $\frac{1}{4}$ of the light intensity. This is called an inverse square law.

The light intensity is **inversely proportional** to the (distance)^2.

This is written as:

$$\text{light intensity} \propto \frac{1}{d^2}$$

Where d is the distance from the lamp.

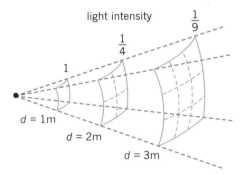

WORKED EXAMPLE

In a photosynthesis experiment, a lamp is positioned 50 cm from a plant and the light intensity on the plant is 10 units. For the second reading, the light is 25 cm from the plant. For the third reading, it is 10 cm from the plant. Write the ratio of the light intensities for the three readings.

Second reading

The lamp is moved from 50 cm to 25 cm, so the distance is halved, or × 0.5.

Light intensity is $\times \dfrac{1}{(0.5)^2} = \times 4$ $10 \times 4 = 40$ units

Third reading

The lamp is moved from 50 cm to 10 cm, so the distance is one fifth, or × 0.2.

Light intensity is $\times \dfrac{1}{(0.2)^2} = \times 25$ $10 \times 25 = 250$ units

Ratio of three readings = 10 : 40 : 250
Divide all by 10.
Ratio = 1 : 4 : 25

PRACTICE QUESTION

5 **B** In a photosynthesis experiment, a lamp is moved so the plant is 1.5 times further away. The light intensity was 8 units. Calculate the new light intensity.

1.13 Proportion and probability

Proportion

Proportion is often used as another word for fraction.

For example, the proportion of the *atoms* in a sample of glucose that are hydrogen atoms is the fraction of hydrogen atoms in a glucose molecule, because every glucose molecule has the same formula: $C_6H_{12}O_6$.

There are $(6 + 12 + 6 =)$ 24 atoms in a glucose molecule and 12 of these are hydrogen.

So the proportion is $\frac{12}{24} = \frac{1}{2}$

You can say the proportion is a half, or write it as a percentage, 50%.

The proportion of the *mass* of glucose that is hydrogen gives a completely different answer because hydrogen atoms have a different mass to oxygen atoms and both have a different mass to carbon atoms.

Probability

The probability, or chance, of something happening can be expressed mathematically.

- If something is certain to happen, the probability = 1.
- If something is impossible then the probability = 0.

Everything else is somewhere in between. If you toss a fair coin, it would be equally likely to land as a head or as a tail. So, the probability of a head = 0.5 or 50% and the probability of a tail = 0.5 or 50%.

✓ WORKED EXAMPLES

1 **What proportion of the mass of glucose is carbon?**

   Formula of glucose = $C_6H_{12}O_6$

   Relative formula mass of glucose (from the periodic table)
   $= (12 \times 6) + (1 \times 12) + (16 \times 6) = 180$

   $\text{Proportion} = \dfrac{\text{relative formula mass of carbon}}{\text{relative formula mass of glucose}} = \dfrac{12}{180} = 6.7\%$

2 **The Punnett square diagram shows the colour outcomes when two chickens breed. State the ratio of black to red chickens, the proportion of black chickens and the probability of a red chicken.**

	A	a
A	AA black	Aa black
a	Aa black	aa red

   The ratio of the colours black to red is 3:1.

   The proportion of black chickens is 3 in every $(3 + 1 =)$ $4 = \dfrac{3}{4}$ or 75%.

   There are four possibilities that are equally likely. So for any one egg, the probability of a red chicken is one out of four $= \dfrac{1}{4}$ or 25%.

PRACTICE QUESTIONS

1 **C** Give the proportion of iron:

   a   by number of atoms

   b   by mass in iron sulfate ($FeSO_4$).

2 **B** In horses, smooth hair is dominant and curly hair is recessive. The genotypes of the parents are Aa and aa. Draw a Punnett square and for the offspring state:

   a   the ratio of smooth hair to curly hair foals

   b   the proportion of smooth to curly hair foals

   c   the probability of a foal with curly hair.

3 **B** Having freckles is a dominant trait. The genotypes of the parents are Aa and Aa. What is the probability of them having a child with freckles?

2 HANDLING DATA

2.1 Significant figures

A student measures a pen with a ruler and tells you it is 14.47 cm long. A good question is, 'How do you know it is 14.47 cm and not 14.48 cm?' The digit 7 is only seven-tenths of a millimetre and you could not measure to this accuracy with a ruler. The 7 is not significant. The pen is between 14.4 cm and 14.5 cm long and closest to 14.5 cm. We can say it is 14.5 cm long to 3 significant figures.

Uncertainties

When you measure something, there will always be a small difference between the measured value and the true value. This may be because of the size of the scale divisions on your measuring equipment, or the difficulty of taking the measurement. Don't confuse this with errors made in measurements – these can be, and should be, avoided. Instead, this is called an uncertainty.

Uncertainty in calculation

The length of a ramp measured with a ruler marked in mm is 300 mm, but at each end of the ruler, there is a 0.5 mm difference between the real value and the measurement. The length could be as much as 301 mm or as little as 299 mm. This uncertainty is written as ± 1 mm.

A drop of oil takes 269 s to flow down the ramp, but the smallest time division on the watch is 1 s. The time could be as much as 270 s or as little as 268 s. There is an uncertainty of ± 1 s.

Calculating the speed of the oil drop when distance $= 300$ mm and time taken $= 269$ s:

$$\text{speed} = \frac{\text{distance travelled}}{\text{time taken}} = \frac{300}{269} = 1.115\,242\ldots \text{mm/s} = 1.1 \text{mm/s rounded to}$$

1 decimal place (see Topic 1.2).

But if the distance was really 301 mm and the time was 268 s:

$$\text{speed} = \frac{\text{distance travelled}}{\text{time taken}} = \frac{301}{268} = 1.123\,134\ldots \text{mm/s} = 1.1 \text{mm/s to 1 d.p.}$$

Or if the distance was 299 mm and the time was 270 s:

$$\text{speed} = \frac{\text{distance travelled}}{\text{time taken}} = \frac{299}{270} = 1.107\,407\ldots \text{mm/s} = 1.1 \text{mm/s to 1 d.p.}$$

This shows you that if you write speed $= 1.115\,242$ mm/s the last 5 digits '15 242' are not significant, because the speed could be anywhere between 1.107 407 mm/s and 1.123 134 m/s.

All three calculations round to 1.1 mm/s. The two digits '1' are significant. The answer is 1.1 mm/s to 2 significant figures.

Identifying significant figures

These numbers are written to 3 significant figures (3 s.f.). The 3 significant figures are underlined:

<u>7.88</u>   <u>25.4</u>   <u>741</u>

Bigger and smaller numbers with 3 significant figures:

0.000<u>147</u>   0.0<u>147</u>   0.<u>245</u>   <u>394</u>00   <u>962</u>00 000

Notice that the zeros before the figures and after the figures are *not* significant – they just show you how large the number is by the position of the decimal point.

Standard form numbers with 3 significant figures:

9.42×10^{-5} 1.56×10^{8}

3 significant figures where the zeros *are* significant:

207   4050   1.01   (any zeros between the other significant figures *are* significant)

85.0   0.760   (normally you would write 85 and 0.76 – the extra zero shows this zero *is* significant)

If the value you wanted to write to 3.s.f. was 590, then to show the zero was significant you would have to write:

590 (to 3.s.f.)   or 5.90×10^{2}

✓ WORKED EXAMPLE

1   **Write the following measurements to 2 s.f.**

   a   634 g = 630 g   **b**   8.65 dm³ = 8.7 dm³

   c   1.97 W = 2.0 W (rounding the number gives 2, but to 2 s.f. this is written 2.0)

2   **Write the following measurements to 3 s.f.**

   a   12.395 m = 12.4 m   **b**   9.051 kg = 9.05 kg

   c $5.0049 \times 10^{6}\,\Omega = 5.00 \times 10^{6}\,\Omega$

? PRACTICE QUESTIONS

1   **B** How many significant figures are there in the following numbers?

   a   45 g   **b**   0.7 cm   **c**   250 kg   **d**   36.9 °C

   e   1650 kJ   **f**   0.0101 mm   **g** 2.00×10^{6} mm³

2   **C** How many significant figures are there in the following numbers?

   a   850 dm³   **b**   4 g/dm³   **c**   0.25 mol/dm³   **d**   477 g

   e 6.022×10^{23} atoms per mole   **f**   100 °C

3   **P** Use the equation:

   $$\text{resistance} = \frac{\text{potential difference}}{\text{current}}$$

   to calculate the resistance of a circuit when the potential difference is 12 V and the current is 1.8 mA. Write your answer in kΩ to 3 s.f.

4   **B** Write the following numbers to a 2 s.f. and b 3 s.f.

   a   7644 g   **b**   27.54 m   **c**   4.3333 g   **d** 5.995×10^{2} cm³

5   **C** Use the equation:

   number of molecules = number of moles $\times\ 6.02 \times 10^{23}$ molecules per mole

   to calculate the number of molecules in 0.5 moles of oxygen. Write your answer in standard form to 3 s.f.

6   **B** Use the equation:

   $$\text{volume} = \text{length} \times \text{width} \times \text{height}$$

   to calculate the volume of a rectangular block 2.5 cm × 5.5 cm × 12.5 cm. Write your answer in cm³ to 3 s.f.

2.2 Mean, median, and mode

Finding the average

The average of a set of data is a number that represents all the data. It is 'in the middle' in some way. The most common average is the mean. The mean is often called the average, but sometimes the mode or median is used, and they can also be called an average.

Mean

The mean of a set of values is calculated by adding up the values and dividing by the number of values.

Mode

The mode is the most common value. Some sets of data have no repeated values, so they have no mode.

Median

The median is the middle value in an ordered list. If there is an even number of values, the median is the mean of the two middle numbers.

The median can be useful if the mean is affected by a very small number of very high values – or a very small number of very low values (see worked example).

Range

The range of a set of data is from the lowest value to the highest value.

Outliers

Outliers, also called anomalous results, are values that are very different to the other results. If you can work out a reason why the value is different – so that you know it is incorrect – you can leave it out when calculating the mean, but there must be a good reason for excluding data and you should state what it is. Circle outliers in tables, and on graphs, and state why you have decided they are outliers.

WORKED EXAMPLES

The heights of some children are given below.

Height (m)	1.51	1.43	1.38	1.55	1.43	1.48	1.49

Calculating the children's mean height

There are 7 children.
Add up their heights:
1.51 m + 1.43 m + 1.38 m + 1.55 m + 1.43 m + 1.48 m + 1.49 m = 10.27 m
Mean height = 10.27 ÷ 7 = 1.4671... = 1.47 m (to 3 s.f.)

Calculating the mode of their heights

Only one child has height 1.51 m, 1.38 m, 1.55 m, 1.48 m, and 1.49 m, but two children have height 1.43 m. So 1.43 m is the mode.

Calculating the median value of their heights

Putting the heights in order: 1.38m 1.43m 1.43m <u>1.48m</u> 1.49m 1.51m 1.55m
The median is the middle number (underlined) = 1.48m

Stating the range of the heights

Smallest = 1.38m and largest = 1.55m, so the range = 1.38m to 1.55m

Outliers

Two more children's heights are added: 1.52m and 2.73m. A student starts to use the new data. However, 2.73m is taller than all adults, so it cannot be correct. The student rings the data and states that it is an outlier as it is not possible for a child to be this tall.

Mean or median?

The table shows the shoe sizes for 11 people:

Shoe size	3	4	4	5	5	5	6	6	7	7	12

Calculate the mean and the median and comment on the difference.
Mean value = (3 + 4 + 4 + 5 + 5 + 5 + 6 + 6 + 7 + 7 + 12) ÷ 11
          = 64 ÷ 11 = 5.8 = 6 (1 s.f.)
Median value = 5
The mean is high because one person has a very large shoe size (size 12). The median, in this case, gives a better indication of the average size of this sample. (The mode is also 5.)

? PRACTICE QUESTIONS

1 **B** A student grows sunflowers in the same conditions and measures their heights. The table shows the results:

Height in m	2.45	2.38	2.65	2.81	2.94	2.28	1.97	2.43	2.63

Calculate **a** the mean **b** the median and **c** state the range.

2 **C** The concentration of nitrogen dioxide in parts per billion (ppb) was measured on a street at hourly intervals. The table shows the results:

Time in h:min	05:00	06:00	07:00	08:00	09:00	10:00	11:00	12:00	13:00	14:00	15:00
NO_2 in ppb	11	13	21	23	29	31	42	41	31	28	23

Calculate **a** the mean **b** the median **c** the mode and **d** state the range.

3 **P** A student records how many units of electricity are used each day. The table shows the results:

Day	Monday	Tuesday	Wednesday	Thursday	Friday	Saturday	Sunday
Units in kW/h	13	13	13	14	14	15	11

Calculate **a** the mean **b** the median **c** the mode and **d** state the range.

2.3 Estimating and sampling

Sampling

When biologists study the environment, they often want to know how many of an organism there are present. This is called the population of the organism. There may be a large number over a large area, so that it is not possible to count them all. The biologists can estimate the population by sampling the area.

A biologist counts the number of organisms in a small area. This is called a sample. The sample area must be similar to the whole area, because if it is different the result will be very inaccurate. Usually several samples are taken and the mean population is calculated (see Topic 2.2). The total estimated population is then calculated by multiplying the sample population by the number of times the sample area fits into the total area.

Quadrats are often used to mark out the sample area. They are square frames with a known area, for example, $1\,m^2$. Figure 1 shows a $1\,m^2$ quadrat and a $0.25\,m^2$ quadrat.

The plants inside the quadrat are counted. Any plants that are more than half in the quadrat are included. Any plants that are less than half in are ignored. In the diagram there are four dandelions in the quadrat.

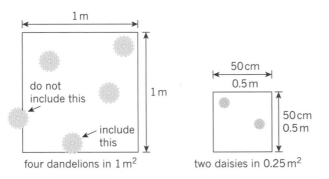

Figure 1 Quadrats showing dandelions and daisies

 WORKED EXAMPLE

A field is 120m long and 75m wide. Some students used $0.25\,m^2$ quadrats to count the number of dandelions in a $0.25\,m^2$ area. They worked out the mean of their results. The mean number of dandelions in a quadrat was 3.5. Estimate the number of dandelions in the field.

You can sketch a diagram to help you:

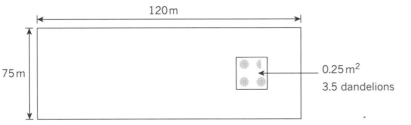

> **NOTE:** Although counting the number in a quadrat gives a whole number – working out the mean may give a number with decimal places.

Area of field = 120m × 75m = 9000 m^2

The number of quadrats that would be needed to cover the field

$$= \frac{\text{area of field}}{\text{area of each quadrat}}$$

$$= \frac{9000 \text{m}^2}{0.25 \text{m}^2} = 36\,000 \text{ quadrats}$$

Number of dandelions = number of quadrats $\times$ mean number of dandelions in a quadrat

Number of dandelions = $36\,000 \times 3.5 = 126\,000$ dandelions

---

PRACTICE QUESTIONS

1 **B** A garden has an area of 100 m². A student uses a 0.25 m² quadrat to count slugs. The mean is 4 slugs in a quadrat. Estimate the number of slugs in the garden.

2 **B** Some students use 1 m² quadrats to count the pieces of plastic waste washed up on a beach. They calculate a mean of 5.2 pieces of plastic waste in the quadrat. The beach is 300 m long and 30 m wide. Estimate the number of pieces of plastic waste on the beach.

3 **B** The diagram shows an investigation by some biologists to estimate the number of wild daffodil plants in a field. Estimate the number of daffodils.

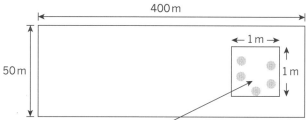

mean number of daffodils = 5

> **REMEMBER:** Sometimes you will need to calculate the area from the length and the width. Sometimes the area is given.
>
> Don't confuse area with length – look at the units:
> - square metres (m²) for measuring area
> - metres (m) for measuring lengths.

STRETCH YOURSELF!

4 **B** A field contains maize plants. The field is 240 m long and 130 m wide.

The farmer uses a quadrat 1.2 m long and 1.2 m wide. He takes 15 random samples. The table shows his results:

Maize plants	8	9	15	10	8	13	8	15	14	9	7	16	8	14	10

Estimate the number of maize plants in the field. Write your answer in standard from to 2 s.f.

2.4 Estimating and orders of magnitude

Estimating calculations

The numbers in calculations can have lots of digits and be complicated to work out. Sometimes it is useful to estimate the answer before calculating it. This can also help you to know if your answer is about the right size. Estimating is not the same as guessing. You may be asked to estimate the result of a calculation in the exam.

To estimate the answer to a calculation, first round all the numbers to 1 s.f. (see Topic 2.1), then work out the value using these numbers.

Your answer must be given to only 1 s.f. As you have rounded the data, it will not be accurate to more than 1 s.f.

✓ WORKED EXAMPLE

Use the equation:

$$speed = \frac{distance}{time}$$

to estimate the speed of a car that travels a distance 1387 m in time 52 s.

1387 m = 1000 m to 1 s.f. 52 s = 50 s to 1 s.f.

Estimated speed $= \dfrac{1000}{50} = 20$ m/s

Actual speed $= \dfrac{1387}{52} = 27$ m/s (2 s.f.)

The estimated speed is quite close to the actual speed.

? PRACTICE QUESTIONS

1. **B** The diameter of the image of a pollen grain is 23.5 mm. It has been magnified × 500 by a microscope. Estimate the actual diameter of the pollen grain. Use the equation:

$$actual\ size\ of\ object = \frac{size\ of\ image}{magnification}$$

2. **C** One mole of calcium hydroxide $Ca(OH)_2$ has a mass of 74 g. Estimate the number of calcium ions in 580 g of calcium hydroxide. Use the equation:

$$number\ of\ ions = number\ of\ moles \times 6.02 \times 10^{23}\ per\ mole$$

3. **P** Estimate the speed of a wave that has frequency 256 Hz and wavelength 1.29 m. Use the equation:

$$wave\ speed = frequency \times wavelength$$

4. **P** Estimate the potential difference across a 4700 Ω resistor when a current of 2.1 mA passes through it. Use the equation:

$$potential\ difference = current \times resistance$$

Orders of magnitude

Orders of magnitude are used to give a general idea of how two objects compare in size. For example, a cherry and a strawberry are similar in volume, so they have the same order of magnitude.

The number of orders of magnitude is the same as the power of 10 (see Topic 1.3).

The order of magnitude of 10 is 1, because $10 = 10^1$, and the order of magnitude of 100 is 2, because $100 = 10^2$.

An apple is about ten times larger than a cherry. They have a difference of one order of magnitude.

A melon is about one hundred times larger than a cherry. They have a difference of two orders of magnitude.

This can be written: the melon is ~100 times bigger than the cherry. The symbol ~ means 'about'.

When comparing sizes, or numbers, divide the large number by the small number.

If the answer is less than 10, then the two numbers have the same order of magnitude.

- $10^0 = 1$ – the same order of magnitude (or zero difference in order of magnitude).
- $10^1 = 10$ – a difference of 1 order of magnitude, one object is about ten times the size of the other.
- $10^2 = 100$ – a difference of 2 orders of magnitude, one object is about a hundred times the size of the other.
- $10^3 = 1000$ – a difference of 3 orders of magnitude, one object is about a thousand times the size of the other.

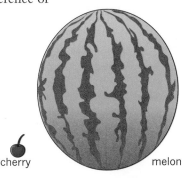

cherry   melon

strawberry

apple

Figure 1 Orders of magnitude

WORKED EXAMPLE

The diameter of a marble is about 2 cm and the diameter of a beach ball is about 28 cm. Compare their sizes using orders of magnitude.

Divide the large number by the small number

$= \dfrac{28}{2} = 14 \quad 10 < 14 < 100$

The beach ball is one order of magnitude larger than the marble.

It is ~10 times bigger than the marble.

PRACTICE QUESTIONS

5 **B** Compare the orders of magnitude of the following objects.

   a a mouse ~80 mm long and an elephant ~6 m long

   b the diameters of a red blood cell ~7 μm and a virus ~160 nm

   c a palisade leaf cell ~70 μm long and a leaf ~7 cm long

6 **C** Compare the orders of magnitude of the following objects.

   a the diameters of a marble ~1 cm and an atom ~0.1 nm

   b the diameters of a proton ~8×10^{-16} m and an electron ~4×10^{-19} m

7 **P** Compare the order of magnitude of the following objects.

   a the mass of a lorry ~44 000 kg and a human body ~70 kg

   b the mass of the lorry and the Earth ~6×10^{24} kg

2.5 Tables

Drawing and interpreting tables

Tables are a useful way of organising and presenting data.

WORKED EXAMPLE

1 **A candle was used to heat water in a beaker. The temperature of the water was measured at regular time intervals. Table 1 shows the results of the experiment.**

Table 1

name of variable

independent variable ——▶

unit

dependent variable

increasing size

Time in minutes	Temperature in °C
0	20
2	40
4	52
6	60
8	67
10	72
12	74

> **REMEMBER:** The **independent variable** is the one you change.
>
> The **dependent variable** is the one that changes because the independent variable changes.
>
> The columns must be labelled with the name of the variable and the units.

a **Determine the time to increase the temperature by 40 °C.**

b **Discuss whether the candle will boil the water.**

a

Notice that you are asked for the time to increase the temperature **by** 40 °C not to 40 °C. A common mistake would be to find the value 40 °C in the table and the corresponding time of 2 minutes.

Calculate the time taken to increase the temperature at the start by 40 °C as follows.

Step 1: find the temperature at the start, 0 minutes, and read

off the temperature = 20 °C.

Step 2: calculate the increase 20 °C + 40 °C = 60 °C.

Step 3: find 60 °C in the table and read off the corresponding time = 6 minutes.

> **NOTE:** When tables have lots of columns use a ruler to line up and read the values.

b

The water will boil at 100 °C. In the first 2 minutes, the temperature increases by 20 °C, but the increases in each subsequent 2 minutes get smaller until after 10 minutes the temperature of the water only increases by 2 °C. Increases will continue to get smaller, so the water temperature will not increase more than another few degrees and so will not boil.

2 Table 2 gives the melting points and boiling points of some chemical elements and compounds.

Table 2

Name	Formula	Melting point in °C	Boiling point in °C
calcium chloride	$CaCl_2$	775	1935
sodium chloride	NaCl	808	1465
sodium hydroxide	NaOH	323	1388
sulfuric acid	H_2SO_4	10	337
water	H_2O	0	100
zinc	Zn	419	906

a List the formulae of the elements and compounds which have:

   i a higher melting point than zinc

   ii a higher boiling point than sodium hydroxide.

b A student says that the higher the melting point, the higher the boiling point. Is this correct? Justify your answer.

a i

Step 1: in Table 2, look for zinc and read across the row for its melting point = 419 °C.

Step 2: start at the top of the column and write down the formula of each element that has a melting point higher than 419 °C. This would be $CaCl_2$ and $NaCl$.

> **NOTE:** Use your ruler to line-up the formula and the melting point and keep track of the row you have reached.

a ii

Step 1: in Table 2, look for sodium hydroxide and read across the row for its boiling point = 1388 °C.

Step 2: start at the top of the column and write down the formula of each element that has a boiling point higher than 1388 °C. This would be $CaCl_2$ and $NaCl$.

> **NOTE:** Use your ruler to line-up the formula and the boiling point and keep track of the row you have reached.

b

Step 1: if you can write on the table, you can number the entries from lowest to highest.

- For melting point: water = 1, sulfuric acid = 2, and so on.
- For boiling point: water = 1, sulfuric acid = 2, and so on.

Otherwise, write out the names or formulae in order from lowest to highest.

- Melting point lowest to highest: H_2O H_2SO_4 $NaOH$ Zn $CaCl_2$ $NaCl$
- Boiling point lowest to highest: H_2O H_2SO_4 Zn $NaOH$ $NaCl$ $CaCl_2$

Step 2: answer the question.
The student is wrong because zinc has a higher melting point than sodium hydroxide but a lower boiling point.
Or
The student is wrong because sodium chloride has a higher melting point than calcium chloride but a lower boing point.
If you spotted that the sodium chloride and calcium chloride showed the student was wrong, you could go straight to step 2. But if the student was right, you would need to write the whole list to justify your answer.

> **NOTE:** You can write both of these – but just one shows the student is wrong and answers the question.

? PRACTICE QUESTION

1 **C** **a** Draw a table to show the results of adding hydrochloric acid to calcium carbonate.

- The loss in mass of large chips of $CaCO_3$ at 20 s was 0.30 g, at 40 s was 0.48 g, at 60 s was 0.59 g, at 80 s was 0.64 g, at 100 s was 0.66 g, and at 120 s was 0.66 g.
- For the small chips of $CaCO_3$ at 20 s was 0.57 g, at 40 s was 0.66 g, at 60 s was 0.68 g, at 80 s was 0.68 g, at 100 s was 0.68 g, and at 120 s was 0.68 g.

   b Describe the effect of changing the chip size on the rate of reaction.

2.6 Bar charts

Discrete and continuous variables

Bar charts are used to display data when the independent variable (the one on the horizontal axis) is **discrete**. This means there are no 'in-between' values.

Examples of discrete data are names, eye colour, and number of offspring.

The dependent variable is **continuous**. This means the numbers can take a range of values.

Examples of continuous data are lengths, weights, heights, and ages.

When the independent variable is continuous the values are split into ranges with no gaps between them, and the chart is called a histogram. For more about histograms, see Topic 2.9.

Bar charts have gaps between the bars. Histograms do not have gaps.

Drawing bar charts

To draw a bar chart, follow these steps.

Step 1: count up the number of bars you need for the independent variable (the horizontal axis or x-axis).

Step 2: choose a bar width so that they will all fit on the axis along the bottom of the paper, allowing for spaces between the bars. Use the same size bar widths, and the same size spaces (the spaces may be smaller than the widths). Make sure you use more than half of the paper.

Step 3: find the largest value of the dependent variable – this will be the tallest bar. Choose a scale so that this fits on the vertical axis (or y-axis) and uses up more than half of the graph paper.

Step 4: with the scale you choose, each square must have a sensible value, for example = 1, 2, 4, 5, or 10. Sometimes each square may be a larger value, for example, 20 or 100. Never chose a value such as 3 or 7. These are very difficult to draw and to read off, and you will lose the mark for choosing a sensible scale.

Step 5: use a ruler to draw the lines and to line up the top of the bars with the scale.

Step 6: give the chart a title, and label the axes with the variable name and the unit.

WORKED EXAMPLES

1 **Draw a bar chart to show the power of the electrical appliances in Table 1.**

Table 1

Electrical appliance	Power rating in W
microwave	1200
fan heater	2000
hairdryer	1800
kettle	3000
oven	2200

The independent variable is the electrical appliance. There are five, so you need five bars. The name of an appliance is a discrete – not a continuous – variable, so allow a space between the bars. Keep the space the same.

The dependent variable is the power rating. The largest value is 3000 W.

A scale of 10 small squares = 1000 W has been chosen.

Use a ruler to draw the axes.

Use a ruler, as shown in Figure 1, to draw the bars to the correct height.

If you are short of time, don't spend time shading the bars – use diagonal lines or another quick way to make the bars look different.

Label the axes, don't forget the units, and give your chart a title.

Figure 1

2 In this bar chart (Figure 2), there are two sets of information for different years.

There is no gap between the bars for the different years, but there is still a gap between the independent variables (the weeds).

This bar chart needs a key for the different years.

Table 2

Weeds	Number counted	
	2016	2017
dandelions	32	28
ragwort	25	12
thistles	30	20

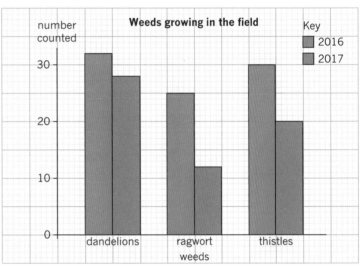

Figure 2

PRACTICE QUESTIONS

1 **B** Use the data below to draw a bar chart of the birds observed at a coastal site.

Birds	Number observed
herring gull	25
black headed gull	18
oystercatcher	13
curlew	4
turnstone	22

2 **C** Use the periodic table to get the data to draw a bar chart for the number of protons in the nucleus of each of these elements: hydrogen, carbon, oxygen, sodium, magnesium, calcium.

3 **P** Draw bar charts for **a** the melting points and **b** the boiling points of the elements and compounds in Table 2 in Topic 2.5.

2.7 Bar charts 2

Interpreting bar charts

Look at the axes carefully to make sure that you know what the bar chart shows.

Use a ruler to line up and read off the values you need.

WORKED EXAMPLE

Figure 1 shows the carbohydrate, protein and fat content of some vegetables.

a **Determine which vegetable has i the highest fat content ii the lowest fat content iii the highest carbohydrate content.**

b Determine i the protein content of peas ii the carbohydrate content of red peppers.

c How much more fat per 100 g is there in peas than in red peppers?

d How much more carbohydrate per 100 g do carrots contain than red peppers?

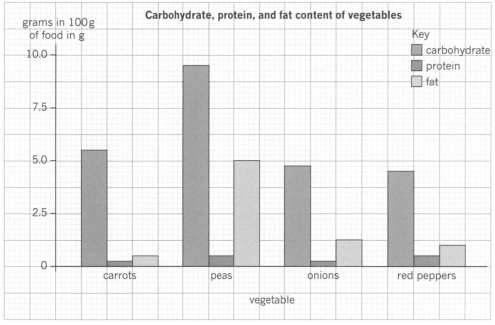

Figure 1

a **i** Use the key to decide which colour bar represents fat.

Look for the tallest bar in that colour = peas.

ii Shortest bar = carrots

iii Use the key to decide which colour bar represents carbohydrate.

Look for the tallest bar in that colour = peas.

b **i** Use a ruler to line up the top of the protein bar for peas with the vertical axis. The bar is 2 small squares.

10 small squares = 2.5

1 small square = 0.25

2 small squares = 0.5, so the protein content = 0.5 g per 100 g.

ii Use a ruler to line up the top of the carbohydrate bar for red peppers with the vertical axis. The bar is 2 small squares *below* 5.0.

5 − 0.5 = 4.5, so the carbohydrate content = 4.5 g per 100 g.

c   Use a ruler to line up the top of the fat bar for peas with the vertical axis. Read off the value on the vertical axis: peas = 5.0 g.

Repeat for the red peppers: red peppers = 1.0 g.

Peas contain 5.0 g – 1.0 g = 4.0 g per 100 g of fat more than red peppers.

d   Use a ruler to line up the top of the carbohydrate bar for carrots with the vertical axis. Read off the value on the vertical axis: carrots = 5.5 g.

Repeat for the red peppers: red peppers = 4.5 g.

Carrots contain 5.5 g – 4.5 g = 1.0 g per 100 g of carbohydrate more than red peppers.

PRACTICE QUESTIONS

1   **P** The chart shows the percentage of electricity generated in the UK by different fuel types.

Percentage of electricity generated by different fuel types

Use the bar chart to determine the answers to the following questions.

a   Which fuel was used to generate most of the electricity in the UK in 2016 and 2017?

b   Which sources of electricity generation i decreased and ii increased from 2016 and 2017?

c   What was the total percentage generated from coal and gas (the fossil fuels) in each year?

2   **B** Figure 2 in Topic 2.6 is a bar chart showing weeds in a field.

Use the bar chart to determine the following information:

a   the most common weed in 2016

b   the number of thistles in 2017

c   the decrease in ragwort plants between 2016 and 2017

d   the difference between the number of dandelions and thistles in i 2016 ii 2017.

STRETCH YOURSELF!

e   Use Figure 2 in Topic 2.6 to calculate the percentage of weeds that were thistles in 2017.

2.8 Frequency tables

Frequency tables

How often something happens is the frequency. This can be recorded in a frequency table or tally chart.

* Sometimes the data falls into separate categories. For example, the days of the week in Table 1.
* Sometimes there is a continuous range of values and they are put into groups. For example, the heights in Table 2.

WORKED EXAMPLES

1 A student counted the number of ripe strawberries from a strawberry field on different days. Their results are shown below.

Table 1

Day	Tally of number of strawberries per day	Number of strawberries per day
Monday	ЖҬ ЖҬ ЖҬ ЖҬ III	23
Tuesday	ЖҬ ЖҬ II	12
Wednesday	ЖҬ ЖҬ ЖҬ ЖҬ ЖҬ I	26
Thursday	ЖҬ ЖҬ ЖҬ III	18
Friday	ЖҬ ЖҬ ЖҬ	15
Saturday	ЖҬ I	6
Sunday	ЖҬ ЖҬ ЖҬ ЖҬ I	21

← frequency

The student counting the strawberries has drawn a tally line for each strawberry. Every fifth tally line is drawn across the group of four to make it easy to count up the total tally to get the frequency.

2 Draw a frequency table of the following height values for a group of students.

Height in cm	164	172	175	148	152	165	194	153	159
	162	183	167	167	176	161	155	169	158

> **REMEMBER:**
> The total of the frequency values = total number of data values. This is a good check that you haven't missed any.

Table 2

Height h in cm	Tally of heights	Students in the height range
$130 < h \leq 150$	I	1
$150 < h \leq 160$	ЖҬ	5
$160 < h \leq 165$	IIII	4
$165 < h \leq 170$	III	3
$170 < h \leq 180$	III	3
$180 < h \leq 200$	II	2

> **REMEMBER:**
> $<$ means 'less than'
> $\leq$ means 'less than or equal to'
> $>$ means 'greater than'
> $\geq$ means 'greater than or equal to'

PRACTICE QUESTIONS

1 **B** The following is a list of shoe sizes for a class of Year 11 students. Draw a frequency table of shoe sizes for the class.

5 7 9 6 4 8 12 8 4 6 5 8 9 3 5 5 7 10 8 4 5 10 6 11 3 6 11 4 9 8 7 9 7 10

2 **B** The following is a list of heart rates for a group of 30 people immediately after exercise.

Heart rate H in beats/ min	148	161	171	143	149	180	158	157	168	150
	153	169	148	152	166	162	150	175	152	155
	156	164	170	168	149	161	179	164	168	176

Draw a frequency table for the following ranges of heart rate.

$140 < H \leq 150$ $150 < H \leq 160$
$160 < H \leq 170$ $170 < H \leq 180$

3 **C** The following measurements of monthly mean nitrogen dioxide pollution levels in µg/m³ were measured at different sites in England.

NO$_2$ in µg/m³	20	47	24	27	40	22	24	17	24	13
	40	17	28	41	33	28	45	27	13	24
	7	25	14	22	10	19	24	35	44	30

Draw a frequency table for the following concentration ranges in µg/m³.

$0 < \text{conc.} \leq 10$ $10 < \text{conc.} \leq 20$ $20 < \text{conc.} \leq 30$
$30 < \text{conc.} \leq 40$ $40 < \text{conc.} \leq 50$

4 **P** The following measurements of car speeds were made by a speed camera on a motorway.

Speed v in m/s	18.0	24.5	30.0	29.0	18.0	23.0	15.0	28.0	26.0	25.0
	26.5	32.0	14.0	35.0	38.0	27.0	31.0	33.5	21.0	30.0
	25.0	29.5	26.0	31.0	30.0	31.5	25.0	30.5	31.0	29.0

Draw a frequency table for the following speed ranges in m/s.

$0 < v \leq 10.0$ $10.0 < v \leq 20.0$ $20.0 < v \leq 25.0$
$25.0 < v \leq 30.0$ $30.0 < v \leq 40.0$

2.9 Histograms

Drawing histograms

Histograms are similar to bar charts. They are used when the independent variable is **continuous**. In a bar chart, the widths of the bars are all the same, but in a histogram, the widths can be different.

Look at the bar charts in Topic 2.6. In Figure 1, the independent variable is electrical appliances, and in Figure 2, it is vegetables. These are **discrete** variables. This means they don't have any in-between values.

For discrete variables, the bars are separated and all have the same width.

In the second worked example below (sunflowers), the independent variable is height. Height is a **continuous variable**.

Look at the frequency tables in Topic 2.8. In Table 2, the independent variable is height and the heights are grouped together so that each bar has a range of values. There are no gaps.

In the first worked example below (bees), the time ranges are all 10 s, so the bars will have the same width and the histogram will look like a bar chart.

WORKED EXAMPLE

Use the information in the following frequency table to draw a histogram for the time bees spend collecting pollen from a single flower.

each of these time ranges is called a group or class

Time t in seconds	$0 < t \le 10$	$10 < t \le 20$	$20 < t \le 30$	$30 < t \le 40$	$40 < t \le 50$
Number of bees	5	18	15	7	0

this is the frequency

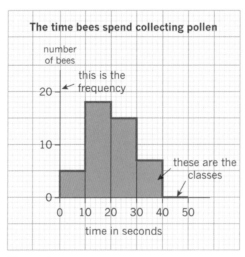

Figure 1

PRACTICE QUESTIONS

1 **B** Draw a histogram to represent the following information about the time students spend doing aerobic exercise each day.

Time t exercising in hours	$0 < t \leq 0.5$	$0.5 < t \leq 1.0$	$1.0 < t \leq 1.5$	$1.5 < t \leq 2.0$	$2.0 < t \leq 2.5$	$2.5 < t \leq 3.0$
Number of students	18	25	28	12	2	0

2 **C** Draw a histogram for the data on nitrogen dioxide pollution levels in practice question 3 in Topic 2.8. (Hint: use the frequency table you drew to answer this question.)

2.10 Histograms 2

WORKED EXAMPLE

Draw a histogram for the following data on the heights of sunflowers in a field.

The first two columns are the frequency table. Add a third and a fourth column for your working out, as shown in Table 1.

Table 1

the height ranges are the classes

this is the frequency

these are for your working out

Height *h* in cm	Number of sunflowers	Class width in cm	Frequency density
$0 < h \le 20$	44	20 – 0 = 20	44 ÷ 20 = 2.2
$20 < h \le 30$	28	30 – 20 = 10	28 ÷ 10 = 2.8
$30 < h \le 40$	26	40 – 30 = 10	26 ÷ 10 = 2.6
$40 < h \le 60$	36	60 – 40 = 20	36 ÷ 20 = 1.8
$60 < h \le 100$	48	100 – 60 = 40	48 ÷ 40 = 1.2

In this, case the ranges are different. The first and fourth are 20 cm, the second and third are 10 cm, and the last one is 40 cm. This means the bars will have different widths and the vertical axis must show the frequency density, not the frequency.

Step 1: add columns to calculate the **class width** and the **frequency density**.

Step 2: calculate the class widths and add them to the table.
The class width is the range of heights. So for the first class, class width = 20 – 0 = 20 cm.

The frequency density $= \dfrac{\text{frequency}}{\text{class width}}$

For the first class, frequency density $= \dfrac{44}{20} = 2.2$.

Step 3: choose sensible scales, just as you would for a bar chart (see Topic 2.6).
The classes go on the horizontal axis. Add them up: 20 + 10 + 10 + 20 + 40 = 100. So 100 must fit on the horizontal axis with no spaces between them.
The frequency density goes on the vertical axis. Find the largest value = 2.8. So 2.8 must fit on the vertical axis.

Step 4: draw the bars using a ruler to get straight lines, and line up values with the axes.
As Figure 2 shows, the bars do not have a gap between them and their widths are equal to the class width.

Step 5: label the axes.

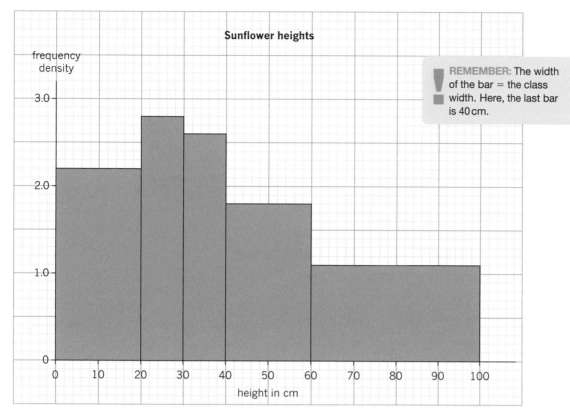

Figure 1

PRACTICE QUESTIONS

1　**B**　Draw a histogram for the data on students' heights in worked example 2 in Topic 2.8.

2　**P**　Draw a histogram for the data on car speeds in practice question 4 in Topic 2.8.

3　**B**　Draw a histogram for the data on heart rates in practice question 2 in Topic 2.8. (Notice that the first class doesn't start at zero.)

2.11 Histograms 3

Histogram questions

When the class widths are all equal in a histogram, and the vertical axis is the **frequency**, calculations are very similar to calculations for bar charts.

WORKED EXAMPLE

The histogram in Figure 1 shows the number of ready-meals with a range of energies in kilojoules. Use it to determine:

a how many meals had between 1500 kJ and 2000 kJ of energy

b how many meals had more than 2500 kJ of energy

c how many meals had between 2000 kJ and 3000 kJ energy

d how many ready-meals were surveyed in total.

Energy contained in ready-meals

number of meals

energy in kJ

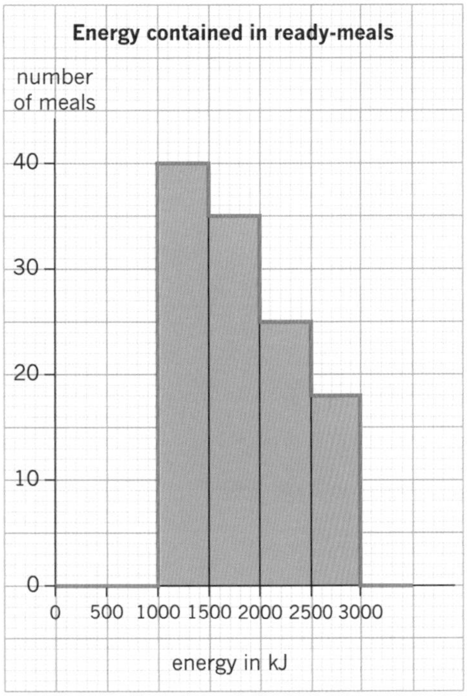

Figure 1

a Use your ruler to line up the bar between 1500 and 2000 kJ of energy. Mark the axis. Read off the value.

1 small square = 1 meal, so there are 30 + 4 = 34 meals.

b There is only one bar higher than 2500 kJ. Line up and do the same as above.

There are 20 − 2 = 18 meals.

c There are two bars between 2000 kJ and 3000 kJ, so the total is the sum of these two bars. Line up and do the same as for part a.

There are 26 + 18 = 44 meals.

> **REMEMBER:** Write down the 26 and the 18. If one of them, or your addition, is incorrect you may still get a mark for reading the graph correctly.

d This is the sum of all the bars. There is only one you have not used so far, 1000 kJ − 1500 kJ. Line up and read this value = 40 − 2 = 38 meals.

The total is:

0 + 0 + 38 + 34 + 26 + 18 + 0 = 116 meals.

PRACTICE QUESTION

1 B Use the histogram showing the time bees spend collecting pollen in Figure 1 in Topic 2.9 to determine:

a how many bees spent between 30 s and 40 s collecting pollen

b how many bees spent less than 20 s collecting pollen

c how many bees were surveyed in total.

Histograms with frequency density on the vertical axis

When the vertical axis of a histogram is frequency density, this value is not usually the answer you want.

The number represented by a bar in this type of histogram is represented by the area of the bar, not the height:

number of occurrences (or frequency) = frequency density × class width

> **REMEMBER:** To calculate the frequency density, you divided the frequency by the class width. So now, to find the frequency, you must multiply the frequency density by the class width.

WORKED EXAMPLE

The histogram in Figure 2 shows the number of newborn babies with a range of mass in kilograms. Use it to determine the following:

a   the number of babies with mass less than 2.0 kg

b   the number of babies with mass in the range 3.0 to 3.5 kg

c   the number of babies with mass under 3.0 kg.

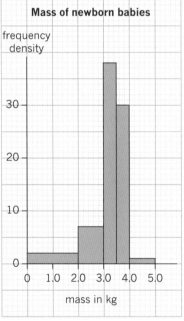

Figure 2

a   There is one bar less than 2.0 kg.

It has height (frequency density) = 2 small squares = 2

It has class width 2.0 − 0 = 2.0 kg

Frequency density × class width = 2 × 2.0 = 4

So there are 4 babies with mass less than 2.0 kg.

b   There is one bar between 3.0 kg and 3.5 kg: height = 38 width = 3.5 − 3.0 = 0.5 kg

Frequency density × class width = 38 × 0.5 = 19

So there are 19 babies with mass between 3.0 kg and 3.5 kg.

c   There are two bars under 3 kg. From part **a** babies with mass less than 2.0 kg = 4.

Babies with mass between 2.0 and 3.0 kg: height = 7 width = 3.0 − 2.0 = 1.0 kg

Frequency density × class width = 7 × 1.0 = 7

The total babies less than 3.0 kg = babies less than 2.0 kg + babies between 2.0 and 3.0 kg = 4 + 7 = 11 babies.

PRACTICE QUESTION

2   B   Use the histogram showing the heights of sunflowers in Figure 1 in Topic 2.10 to determine:

a   how many sunflowers were between 40 and 60 cm tall

b   how many sunflowers were over 60 cm tall

c   how many sunflowers were less than 30 cm tall

d   how many sunflowers were there in total.

2.12 Correlation and scatter diagrams

Correlation

When there is a link between two variables, so that if one increases the other increases, then you can say there is a **positive correlation** between the two variables.

When there is a link so that if one increases the other decreases, then you can say there is a **negative correlation** between the two variables.

It is tempting to think that this means one variable is affecting the other, but this is not always true.

You may have heard that when sales of ice cream increase there are more cases of sunburn. This is true, but it is *not* because ice cream *causes* sunburn. It is because both increase in hot, sunny weather.

Sometimes, when you measure two things that are completely independent of each other, the results may show a correlation quite by chance.

Scientists are careful to look for a reason to explain any correlation they find, and to repeat the measurements several times before concluding that there is a link.

Scatter diagrams

You can plot a scatter diagram to see whether there is a correlation. This is a graph with one variable on the vertical axis and the other on the horizontal axis. Points are plotted on the graph. A perfect correlation results in a straight line through the points, whereas a weaker correlation results in a scatter of points as if around a straight line. When there is no correlation, the points are random.

Note that some relationships are not linear, so a correlation doesn't have to be a straight line. For example, the braking distance of a car depends on the speed squared, so the plotted points result in a curve.

WORKED EXAMPLE

Identify whether there is a correlation between the variables in each of the following scatter diagrams.

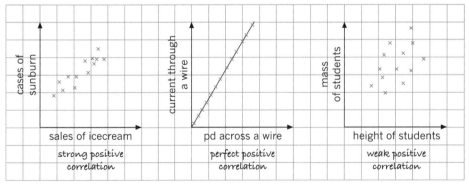

Figure 1 Types of positive correlation

The scatter diagrams in Figure 1 all show positive correlation.

- The first is a strong positive correlation – cases of sunburn increase when sales of ice cream increase.

- The second is a perfect positive correlation. In fact, the current through a wire is directly proportional to the potential difference across it. See Topics 1.11 and 4.3 for more about directly proportional relationships.

- The third is a weak positive correlation – generally taller students have greater mass, but body shape varies by a large amount, and bone density and muscle mass affect the result. So, for this group of students, this was only a weak correlation.

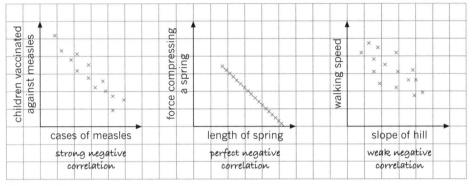

Figure 2 Types of negative correlation

The scatter diagrams in Figure 2 all show negative correlation.

- The first is a strong negative correlation. Measles spreads very easily, so as the number of children vaccinated increases, the number of cases of measles decreases.

- The second is a perfect negative correlation. The spring starts from its natural length and then, as the squashing force is applied, the spring gets shorter. When all the coils are squashed together, the spring will be harder to squash – but this diagram stops before that point is reached.

- The third is a weak negative correlation. As the slope of a hill increases, the speed of the people walking will decrease, but their speed will also depend on their fitness and whether the path is rocky or muddy. So, this is only a weak correlation.

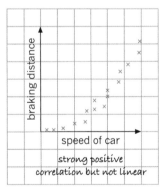
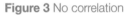

Figure 3 No correlation

Figure 3 shows no correlation. Notice that a random distribution does not mean that all of the points are evenly spaced! It is very difficult to draw a truly random collection of points, but that is what you would expect if you plot two completely unconnected variables. On any one day, the number of babies born in a hospital and the number of birds visiting a bird feeder should not have any connection with each other.

Figure 4 shows a positive correlation between the points, but they lie on a curve, not a straight line. Lots of relationships between variables are not linear. There are many examples, but a few are pressure and volume of a gas, reaction rates of chemicals, radioactive decay and how the intensity of light depends on the distance from a point light source. You can find out more about these in Chapter 4.

Figure 4 A correlation which gives a curve

PRACTICE QUESTION

See Topic 2.13 about drawing scatter diagrams and line graphs. When you have drawn each scatter diagram, decide whether there is a correlation shown, and if so, what type of correlation.

2.13 Drawing scatter diagrams and line graphs

Scatter diagram and line graph FAQs

1 What is the difference between a scatter diagram and a line graph?

You can plot a scatter diagram whether there is a correlation between the variables or not. For an example, see Figure 3 in Topic 2.11. Once you have plotted the points, you may see that you can draw a line through the points. So if there is a strong or perfect correlation there isn't any difference between a scatter diagram and a line graph, but you shouldn't assume you can plot a line graph until you have plotted the points – unless you are told it is a line graph.

2 What do I do if it says draw the best line and it is a curve?

A line can be a straight line or a curve. A linear graph or relationship means a straight line. Be very careful to make sure you haven't plotted a point incorrectly if it does not lie on the straight line of the others.

How to draw a scatter diagram or line graph

You have the graph paper and the data.

Step 1: you will need:

- a transparent ruler marked in cm and mm (a 30 cm one is better than 15 cm; make sure it hasn't got dents in from hitting things, otherwise the line won't be straight)

- two sharp HB pencils (or 1 pencil and a sharpener)

- an eraser.

Step 2: decide which variable goes on the horizontal or x-axis, and which goes on the vertical or y-axis. You may be told this in the question. If not, then the independent variable (the one you change) goes on the x-axis and the dependent variable goes on the y-axis.

Step 3: decide whether you need to include the zero on the graph for each axis. Sometimes it is important either because the line goes through it, or to see if the line goes through it. (For example, if the numbers range from 520 to 645, then you would have to squash the points together to fit 0 to 645 on the graph, so you are probably not supposed to include it.)

Step 4: look for the largest number to help you decide the best scale. When choosing a scale:

- all the points must fit on the graph paper

- the points should take up more than half of the graph paper – for each axis

- the scale should be based on 1, 2, or 5, or multiples of those numbers. For example: 1 square = 0.2 or 1 square = 50. Never choose 3 or 7 they are hard to plot, hard to read, and not sensible, so you will lose marks.

Step 5: mark the values on the axes.

Step 6: find the first x data value on the x-axis and then go up the grid lines until you reach the level of the y data value on the y-axis. Use your ruler to help line up with the axis. Mark the point with a small cross so the centre of the cross is the actual point. Repeat for all the data values.

Step 7: look at your points. Are there one or two that are not in line with the others? If so, check you have plotted them correctly.

The best line through the points (line of best fit)

Step 8: be guided by the question. It may tell you to plot a straight line or a curve.

Step 9: put your ruler through the points. Move it around and see whether the line can be drawn through the origin (0,0) and still be a best line through the points. If you have not been told, decide whether the points form a curve.

Never join one point to another in a series of small straight lines – always draw one line. This is why it is important to move the ruler around aiming for:

- as many points as possible on the line
- the same number of points above and below the line.

If the line starts linear and then curves, be careful not to have a sharp corner where the two lines join. Your curve should be smooth.

Once you have the ruler in the right place, complete steps 10 and 11.

Step 10: with both hands on the ruler, slide it very slightly away from where you want the line, because your pencil line will be above the edge of the ruler.

Step 11: hold the ruler firmly and draw one line. If the graph is a curve, then stop before you get to the curved part and draw the last bit of the straight line and the curve freehand to get it smooth. This must still be one line.

Step 12: label the axes with the name of the variable and the units.

Step 13: you may need to add a title.

You can see examples of scatter diagrams in Topic 2.12 and line graphs in Chapter 4.

PRACTICE QUESTIONS

1 For each of the following tables of data:

   a Plot a scatter diagram

   b Draw a line of best fit

   c Describe the correlation.

Table 1 The effect of varying light intensity on the rate of photosynthesis of a plant.

Light intensity in arbitrary units	Rate of photosynthesis in mm^3/s of CO_2
0	0
0.5	30
1.0	50
1.5	64
2.0	72
2.5	76
3.0	80
3.5	80
4.0	82
4.5	78
5.0	80
5.5	80

Table 2 Concentration of HCl over the course of a reaction.

Time in seconds	Concentration of HCl in mol/dm^3
0	0
25	1.25
50	0.95
75	0.80
100	0.70
125	0.60
150	0.55
175	0.55
200	0.50
225	0.50
250	0.50
275	0.50

1 Calculate the following values. Give your answers using indices.

    a $10^7 \times 10^2$

    b $10^{-3} \times 10^8$

    c $10^5 \div 10^2$

    d $20^6 + 10^{-3}$

2 The average mass of oxygen produced by an oak tree is 11800 g per year. Give this mass in standard form and quote your answer to 2 significant figures.

3 Write the following numbers to **i** 2.s.f and **ii** 3 s.f.

    a 2.008×10^6 m/s

    b 1.602×10^{-19} m/s

    c $3815\,\Omega$

4 Give:

    a 780 km/h in m/s

    b 19 km/h in m/s

    c $50\,cm^3$ in dm^3

5 **a** One in 17 people in the UK has diabetes.

       Calculate the percentage of the UK population that have diabetes.

    b A 500 g sample of a powder is found to be 21% sodium chloride.

       Calculate the mass of sodium chloride that is in the powder.

    c A train accelerates from 55 km/h to 126 km/h.

       Calculate the percentage change in the train's speed.

    d A population of mountain gorillas increased from 480 gorillas in 2010, to 604 gorillas in 2018.

       Calculate the percentage change in the gorilla population.

6 A section of DNA has 800000 bases, of which 31% are T. Calculate the number of each of the four bases: T, A, C, and G. (Remember: A pairs with T and C pairs with G.)

7 Compare the orders of magnitude of:

    a a human hair diameter (100 μm) and a human body (2 m tall)

    b the diameters of an atom (1×10^{-10} m) and the nucleus of the atom ($1 \times 10{-14}$ m)

    c the mass of the Earth (6×10^{24} kg) and the Sun (2×10^{30} kg).

8 A group of students are each given a sunflower seed. They grow the seeds and measure their heights. Their results are shown below.

Height in m	2.37	2.56	0.02	2.95	0.95	1.53	2.04

Calculate:

a the mean

b the median

c the range.

9 Draw a bar chart for the number of ripe strawberries each day shown below.

Day	Number of strawberries per day
Monday	23
Tuesday	12
Wednesday	26
Thursday	18
Friday	15
Saturday	6
Sunday	21

10 For the data in the table below:

a plot a scatter diagram

b draw a line of best fit

c describe the correlation shown.

Current in mA	pd in V
4.0	1.6
5.0	1.9
6.0	2.4
7.0	2.9
8.0	3.2
9.0	3.5
10.0	4.1
11.0	4.3
12.0	4.7
13.0	5.2
14.0	5.4
15.0	6.0

3 EQUATIONS

3.1 Using equations

For GCSE, there are some equations you must learn. You may be tested on the following:

- Do you know the equation?
- Can you use the equation?
- Do you know **and can** you use the equation?

> **REMEMBER:** Learn the equations!
> If you don't know one, and it is not given to you, you can't score marks for:
> - remembering the equation
> - rearranging the equation
> - substituting in the equation
> - calculating the answer.
>
> That is a lot of marks to lose.

There is a list of equations at the back of this book (see pages 116–118). Check which ones you need to know from your exam specification and learn them.

How memory works

You will find that, the first time you learn them well enough to remember them, a few weeks later you have forgotten a lot of them.

Keep coming back to learn them again during your course. Each time you will learn them more quickly, and remember more, until eventually you won't forget them.

The steps when using equations

Step 1: read the question. Don't worry if you don't understand it yet. As you read it, mark the values you are given and the one you want to calculate. If it is your paper, underline, highlight, or write the symbol of the quantities (for example, m for mass or t for time). If it is a book, then write the values with the units on your paper. Draw a rough diagram if it helps.

Step 2: look at what you have written down. You may now realise you do understand the question after all. Decide the equation you need to use and write it down. Don't rearrange it or put values in it yet – just write it down and you may score a mark for remembering it.

Step 3: check if any units need converting. If so, convert the units (see Topics 1.5 and 1.6).

Step 4: if you need to rearrange the equation, you may rearrange the equation first or you may substitute the numbers first. See Topic 3.3 for help with this.

Step 5: substitute the numbers.

Step 6: calculate the answer.

Step 7: check the units of you answer.

Step 8: check whether your answer needs to have a certain number of decimal places (see Topic 1.1) or significant figures (see Topic 2.1).

WORKED EXAMPLES

Example 1

A cheek cell has a $\overset{O}{\underset{\textstyle}{\widehat{0.06\,mm}}}$ diameter. Under a microscope it has a diameter $\underset{I}{\widehat{12\,mm}}$.

What is the $\widehat{\text{magnification}}$?
$$M?$$

Step 1: read and mark the values. This has been already been done above in this example.

Step 2: write down the equation:
$$\text{magnification} = \frac{\text{image size (mm)}}{\text{object size (mm)}} \quad \text{or} \quad M = \frac{I}{O}$$

Step 3: units are the same (mm), so no conversion is needed.

Step 4: you want to calculate magnification M, so no rearrangement of the equation is needed.

Step 5: substitute the values:
$$M = \frac{12\,mm}{0.06\,mm} = \frac{12}{0.06}$$

Step 6: calculate the answer:
$$\frac{12}{0.06} = 200$$

If your magnification is smaller when it should be bigger, you probably have I and O the wrong way round.

Step 7: magnification has no units.

Step 8: there is no need to write the answer to a number of s.f. or d.p.

Answer: magnification $= \times 200$

Example 2

A series circuit is switched on for 10 s. The current through a 4.7 kΩ resistor is 3.5 mA. Calculate the potential difference across the resistor. Give your answer to 2 s.f.

Step 1: $t = 10\,s$ $R = 4.7\,k\Omega$ $I = 35\,mA$ $V = ?$ (to 2 s.f.)

Step 2: pd (or voltage) (V) = current (A) × resistance (Ω)   or $V = IR$

The time t is not needed for this part of the question.

Step 3: change from kΩ and mA:
$$R = 4.7 \times 1000\,\Omega = 4700\,\Omega \qquad I = 3.5 \times \frac{1}{1000}\,A = 0.0035\,A$$

Step 4: is not needed.

Step 5: $V = 0.0035 \times 4700$

Step 6: $V = 16.45$

Step 7: $V = 16.45\,V$

Step 8: $V = 16\,V$ (to 2 s.f.)

PRACTICE QUESTIONS

Choose equations from the lists of equations on pages 116–118.

1 **C** Calculate the concentration in g/dm³ of a solution made by dissolving 52 g of sodium chloride in 1.2 dm³ of water.

2 **P** Calculate the kinetic energy of a car of mass 1200 kg travelling at 28 m/s. Write your answer to 2 s.f.

3 **B** Calculate the magnification of a hair that has a width of 6.6 mm on a photograph. The hair is 165 μm wide.

4 **C** A solution is made by dissolving 0.05 moles of sodium bicarbonate in 25 cm³ of water. Calculate the concentration in mol/dm³.

3.2 More equations

Unpicking the question

Sometimes you are given more data in the question than you need, or the data has to be determined from other data – such as other values or a graph or table.

Think about the data you have. Write down the equation and then think about how to get the data to substitute into the equation.

WORKED EXAMPLE

A firework is let off 495 m from a cliff. A phone at the launch site records the time of the firework at 10:23:05 and the echo from the cliff at 10:23:08. Calculate the speed of sound.

Step 1: you may find it helpful to draw a sketch (Figure 1) to show the data:

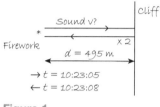

Figure 1

Step 2: speed (m/s) = $\dfrac{\text{distance (m)}}{\text{time (s)}}$ or $v = \dfrac{s}{t}$

Notice that the sound travels to the cliff *and back again*, so the distance is twice the distance from the firework to the cliff. It is a common mistake to forget this.

$s = 495 \times 2 = 990\,\text{m}$ $t = 10:23:08 - 10:23:05 = 3\,\text{s}$

Steps 3 and 4: are not needed.

Step 5: $v = \dfrac{990}{3}$

Steps 6 and 7: $v = 330\,\text{m/s}$

Step 8: is not needed.

Speed of sound $= 330\,\text{m/s}$

PRACTICE QUESTIONS

Choose equations from the lists of equations on pages 116–118.

1   **B** A woman has a mass of 79 kg and she is 168 cm tall. She loses 11 kg. Calculate her new BMI to 2 s.f.

2   **C** A solution is made by dissolving 0.01 moles of calcium chloride in 25 cm^3 of water. Calculate the concentration in **a** mol/dm^3 and **b** g/dm^3.
(Relative formula mass of calcium chloride = 111)

3   **P** Calculate the energy required to heat 70 g of water from 20 °C to 80 °C.
(Specific heat capacity of water = 4200 J/kg °C.)

WORKED EXAMPLE

A piece of metal is put into a measuring cylinder of water on a top pan balance. Before the metal is added, the mass is 113 g and the water level in the cylinder is 74 cm³. Afterwards, the mass is 144 g and the water level is 77.5 cm³. Calculate the density of the metal in kg/m³. Give your answer to 2 s.f.

Step 1: you may find it helpful to draw a sketch (Figure 2) to show the data:

Step 2: density (kg/m³) = $\dfrac{\text{mass (kg)}}{\text{volume (m}^3)}$ or $D \text{ (or } \rho) = \dfrac{m}{V}$

The mass and volume you want to find is that of the metal – not that of the cylinder, water, and metal.

$$\begin{array}{ccc}\text{Mass of} \\ \text{metal}\end{array} = \begin{array}{c}\text{mass of cylinder and} \\ \text{water with metal}\end{array} - \begin{array}{c}\text{mass of cylinder} \\ \text{and water}\end{array}$$

= 144 g − 113 g = 31 g

$$\begin{array}{c}\text{Volume} \\ \text{of metal}\end{array} = \begin{array}{c}\text{volume of water} \\ \text{with metal}\end{array} - \begin{array}{c}\text{volume of water} \\ \text{without metal}\end{array}$$

= 77.5 cm³ − 74 cm³ = 3.5 cm³

Step 3: 31 g = 31 × 10⁻³ kg
1 cm³ = 1 × 10⁻² × 10⁻² × 10⁻² m³ = 10⁻⁶ m³
So 3.5 cm³ = 3.5 × 10⁻⁶ m³

Step 4: is not needed.

Step 5: density = $\dfrac{31 \times 10^{-3}}{3.5 \times 10^{-6}}$

Steps 6 and 7: density = 8857 kg/m³

Step 8: density = 8600 kg/m³ (to 2 s.f.)

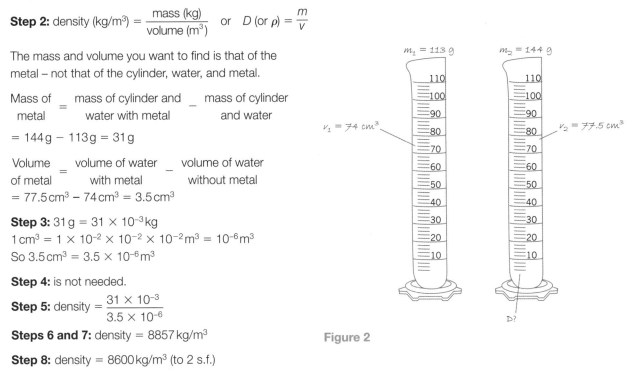

Figure 2

PRACTICE QUESTION

Choose an equation from the lists of equations on pages 116–118.

4 **P** A stone of mass 160 g falls from a cliff 94 m high. Calculate its kinetic energy when it hits the ground. (Gravitational field strength = 10 N/kg)

STRETCH YOURSELF!

Choose equations from the lists of equations on pages 116–118.

5 **P** A weight of 2.2 N is suspended from a long spring. The spring stretches from 0.71 m to 0.79 m. Calculate **a** the spring constant of the spring and **b** the potential energy stored in the spring.

6 **P** A wire 53 cm long is in a magnetic field. The magnetic field is at right angles to the wire and has a magnetic flux density of 0.3 T. A current of 950 mA passes through the wire. Calculate the force on the wire.

3.3 Rearranging equations

Rearranging equations

Sometimes you will need to rearrange an equation to calculate the answer to a question. For example, if you want to calculate the resistance R, the equation:

potential difference (V) = current (A) × resistance (Ω)   or $V = I R$

must be rearranged to make R the subject of the equation:

$$R = \frac{V}{I}$$

There are several ways to do this. Choose the method that works best for you.

Method 1

Think about the quantities and units and use common sense to work out which factor would make the answer bigger (multiply by this) and which would make it smaller (divide by this).

Method 2

Use a triangle. This method needs a good memory. You must remember the triangle and how to use it correctly as you are not using an understanding of what you are doing. If you get the answer wrong, you may not get the mark for a correct equation even if your triangle is correct as the triangle is not an equation. So, you should write the equation as well.

Method 3

Substitute the data into the equation. Then rearrange the equation.

Method 4

Use algebra that you have learned in Maths. Then substitute into your rearranged equation.

> **!** REMEMBER: Whatever you do to one side of the equation you must do the same to the other, so the two sides remain equal (see Topic 3.4).

WORKED EXAMPLES

Method 1

1 Calculate the (time) it takes for a car to travel (480 km) at (50 km/h). Give your answer in (hours).

distance   *speed*

Step 1: read and mark the values as above (see Topic 3.1 for the steps).

Step 2: if a car goes faster (say, 60 mph instead of 30 mph), it will go further – a bigger distance – in the same time. It will take less time to go the same distance.

$$\text{speed (km/h)} = \frac{\text{distance (km)}}{\text{time (h)}}$$

Step 3: is not needed.

Step 4: the time a journey takes will be more if the distance is further, but less if the speed is greater, so:

$$\text{time} = \frac{\text{distance}}{\text{speed}}$$

Steps 5 to 8: $\text{time} = \dfrac{480\,\text{km}}{50\,\text{km/h}} = 9.6\ \text{hours}$

2 Calculate the (distance) a horse gallops in (1 minute) at (12 m/s.)

time   *speed*

The distance the horse travels will be more if the speed is greater and more if the time is greater, so:

$$\text{distance (m)} = \text{speed (m/s)} \times \text{time (s)}$$

1 minute = 60 s

Distance = $12 \times 60 = 72$ m

Method 2

3 Calculate the (actual size) of an amoeba which has a size of (28 mm) when seen with a microscope with magnification (× 200). Give your answer in mm.

0? μm   I   M

Step 1: read and mark the values as above.

Step 2: $\text{magnification} = \dfrac{\text{image size (mm)}}{\text{object size (mm)}}$ or $M = \dfrac{I}{O}$

Step 3: the image is in mm, but the answer (object) is also to be in mm. Magnification has no units. So there is no need to change the units. Keep them in mm.

Step 4: rearrange the equation using a triangle (Figure 1). In this equation, the image is 'over' the object. Put the image 'over' the object in your triangle, and then put the magnification in the remaining space. Or you may have learned this triangle. If so, just write it down.

If you cover M with your finger you can see that $M = \dfrac{I}{O}$, so you have the triangle correct.

Cover the letter representing the answer you want. For example, to find the image size, cover I with your finger. You can see that $I = M \times O$.

In this question, you need to calculate O. Cover O with your finger.

You can see that $O = \dfrac{I}{M}$

Step 5: substitute into the equation:

$O = \dfrac{28}{200}$

Step 6: calculate the answer

$O = 0.14$

Step 7: add units

$O = 0.14$ mm

Step 8: is not needed.

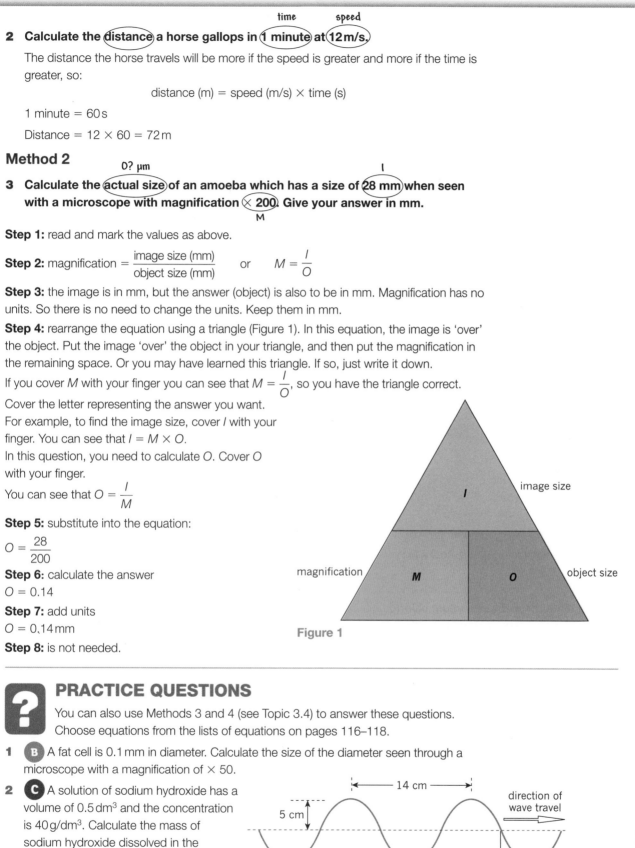

Figure 1

? PRACTICE QUESTIONS

You can also use Methods 3 and 4 (see Topic 3.4) to answer these questions. Choose equations from the lists of equations on pages 116–118.

1 **B** A fat cell is 0.1 mm in diameter. Calculate the size of the diameter seen through a microscope with a magnification of × 50.

2 **C** A solution of sodium hydroxide has a volume of 0.5 dm³ and the concentration is 40 g/dm³. Calculate the mass of sodium hydroxide dissolved in the solution.

3 **P** The figure shows water waves. Use the information on the diagram to calculate the speed of the waves in cm/s.

← 14 cm →

5 cm

direction of wave travel

time between crests passing this point = 8.2 s

3.4 Rearranging equations 2

Rearranging equations

In Topic 3.3, you saw how to rearrange equations by reasoning (Method 1) or memorising a triangle (Method 2). This topic looks at rearranging equations by substituting and then rearranging (Method 3) or by rearranging the equation and then substituting (Method 4).

WORKED EXAMPLES

Method 3

1 A car speeds up from 9 m/s to 20 m/s over a distance of 150 m. Calculate the acceleration to 2 s.f.

Step 1: initial velocity = 9 m/s   final velocity = 20 m/s   distance = 150 m
acceleration = ?

Step 2: (final velocity)^2 – (initial velocity)^2 = 2 $\times$ acceleration $\times$ distance

In this method, go to **Step 5**.

Substitute the values: $20^2 – 9^2 = 2 \times$ acceleration $\times$ 150

Do as much of the calculation as you can:

$400 – 81 = 300 \times$ acceleration

$319 = 300 \times$ acceleration

Now do **Step 4:** rearrange.

The acceleration is multiplied by 300. So to get the acceleration, divide both sides of the equation by 300:

$$\frac{319}{300} = \frac{300}{300} \times \text{acceleration}$$

Step 6: 1.06 = 1 $\times$ acceleration

Step 7: acceleration = 1.06 m/s^2

Step 8: acceleration = 1.1 m/s^2 (to 2 s.f.)

> **REMEMBER:** The steps refer to the steps described in Topic 3.1.

Method 4

2 Calculate the mass of sodium chloride needed to make 75 cm^3 of a 160 g/dm^3 solution of sodium chloride.

Step 1: mass of solute = ?   solvent = 75 cm^3 conc = 160 g/dm^3

Step 2: conc. $= \dfrac{\text{mass of solute}}{\text{volume of solvent}}$

Step 3: 75 cm$^3 = \dfrac{75}{1000}$ dm^3 = 0.075 dm^3

Step 4: rearrange equation.

You want the mass of the solute and it is divided by the volume of solvent. So multiply both sides of the equation by the volume of the solvent:

conc. $\times$ volume of solvent $= \dfrac{\text{mass of solute}}{\text{volume of solvent}} \times$ volume of solvent

$\dfrac{\text{volume of solvent}}{\text{volume of solvent}} = 1$, so the equation is now:

conc. $\times$ volume of solvent = mass of solute

Step 5: 160 $\times$ 0.075 = mass of solute

Steps 6, 7, and 8: mass of solute = 12 g

STRETCH YOURSELF!

3 A 57 kg block falls from a height of 68 m. By considering the energy transferred, calculate its speed when it reaches the ground. (Gravitational field strength = 10 N/kg)

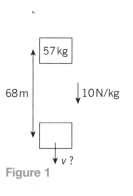

Figure 1

Step 1: $m = 57$ kg $h = 68$ m $g = 10$ N/kg $v = ?$

Step 2: there are three equations:

$$PE = m g h \qquad KE \text{ gained} = PE \text{ lost} \qquad KE = 0.5 m v^2$$

Step 3: is not needed.

Step 4: rearrange the equations.

As KE gained = PE lost, $m g h = 0.5 m v^2$

You want to find v. Divide both sides of the equation by $0.5 m$:

$$\frac{mgh}{0.5m} = \frac{0.5mv^2}{0.5m}$$

$$2 g h = v^2$$

To get v, take the square root of both sides:

$$v = \sqrt{(2 g h)}$$

Step 5: substitute into the equation:

$$v = \sqrt{(2 \times 10 \times 68)}$$

Steps 6, 7, and 8: $v = \sqrt{1360} = 37$ m/s

Notice that in this method you do not need to use the mass.

PRACTICE QUESTIONS

You can also use Method 2 (see Topic 3.3) to answer these questions. Choose equations from the lists of equations on pages 116–118.

1 **B** A Petri dish shows a circular colony of bacteria with a cross-sectional area of 5.3 cm². Calculate the radius of this area.

2 **B** The size of a cell when it is viewed through a microscope at magnification × 500 is 35 mm. Calculate the actual size of the cell.

3 **P** The potential difference across a resistor is 12 V and the current through it is 0.25 A. Calculate its resistance.

4 **C** A 25 g/dm³ solution of sodium bicarbonate was made with 48 g of the solid. Calculate the amount of solvent used.

5 **C** A helium balloon contains 1.3 dm³ of helium gas at room temperature and pressure. Calculate the number of moles of helium in the balloon.

STRETCH YOURSELF!

6 **P** Calculate the specific latent heat of fusion for water from this data:
4.03×10^4 J of energy melted 120 g of ice.
Give your answer in J/kg in standard form.

7 **P** Red light has a wavelength of 650 nm. Calculate its frequency.
Write your answer in standard form.
(Speed of light $= 3.0 \times 10^8$ m/s)

4 GRAPHS

4.1 $y = mx$ and slopes

Straight line graphs

You will probably have seen graphs like this in maths:

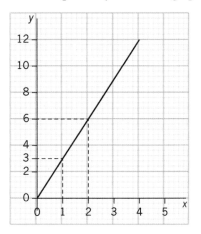

Figure 1 Graph of $y = 3x$

$y = mx$

Figure 1 has dependent variable y and independent variable x. When $x = 1$, $y = 3$ and when $x = 2$, $y = 6$. You can see that $y = 3x$ whatever value of x you choose.

The slope, or gradient, of the straight line is 3. Figure 1 is the graph of the equation $y = 3x$. Notice that when $x = 0$, $y = 0$, so the straight line goes through the origin $(0,0)$.

You can draw a graph with a different slope, for example $y = 5x$ or $y = -19.2x$, so the value of the slope is sometimes given the symbol m.

In Figure 1, $m = 3$. When $y = 5x$, $m = 5$ and when $y = -19.2x$, $m = -19.2$.

A straight line through the origin has the equation $y = mx$, where m is the slope and has a constant value.

The slope of a straight line graph

The slope of the straight line on a graph $= \dfrac{\text{change in } y}{\text{change in } x}$

> **REMEMBER:** A steeper slope is a larger slope. To make the slope steeper, the y value must be bigger. So to make the value larger, y must be on the top of the equation.

Figure 2 shows how to determine the slope of a straight line.

Step 1: choose two points on the line that:

* make a big triangle – the bigger the triangle the more accurate your value will be

* are easy to read off the x- and y-axes, not half way through squares on the graph paper.

Step 2: read the two values on the *x*-axis and subtract them to get the change in *x*. Use a ruler to help you.

Change in $x = 5 - 1 = 4$

Step 3: read the two values on the *y*-axis and subtract them to get the change in *y*. Use a ruler to help you. If the graph is sloping up, this will be a positive value. If the graph is sloping down, it will be a negative value because *y* is getting smaller.

Change in $y = 35 - 7 = 28$

Step 4: calculate: $\dfrac{\text{change in } y}{\text{change in } x} = \dfrac{28}{4} = 7$

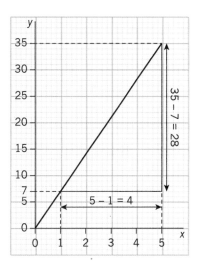

Figure 2 The slope of a straight line

 WORKED EXAMPLE

A science research laboratory uses high voltages (potential differences, pds) for experiments. Figure 3 shows the pd across a component plotted against the current through it. Determine the resistance of the component.

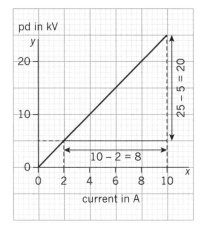

Figure 3

> **REMEMBER:** You can use (0,0) to calculate the gradient if the line goes through (0,0). This example uses the point (2,5) to show you what to do if the line doesn't go through (0,0).
>
> Using (0,0), slope $= \dfrac{25 - 0}{10 - 0} = \dfrac{25}{10} = 2.5$

slope $= \dfrac{20}{8} = 2.5$

This is a number with no units, but you want the resistance so you need to take into account the units:

resistance $= \dfrac{20\,\text{kV}}{8\,\text{A}} = \dfrac{20 \times 1000\,\text{V}}{8\,\text{A}} = 2500\,\Omega$

 PRACTICE QUESTIONS

See Topics 4.2, 4.3 and 4.9 for further questions.

4.2 $y = mx + c$

A straight line graph not through the origin

Figure 1 shows a straight line graph that does not go through the origin.

When $x = 0$, $y = 2$. This value of y is sometimes called 'the y intercept' and is given the symbol c.

The slope of this graph is 3 and you can see that the line is at the same angle as the line in Figure 1 in Topic 4.1.

When $x = 1$, $y = 5$ ($y = 3 + 2$) and when $x = 2$, $y = 8$ ($y = 6 + 2$).

This is the graph of the equation $y = 3x + 2$.

A straight line *not* through the origin has the equation $y = mx + c$, where m is the slope and has a constant value, and c is the value of y when $x = 0$.

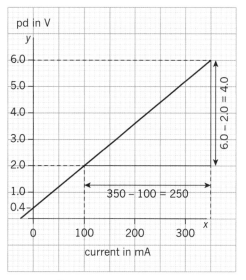

Figure 1 Graph of $y = mx + c$

WORKED EXAMPLE

A student varies the current through a resistor and measures the potential difference (the pd) across it. The voltmeter has zero error and does not read zero when the pd is zero.

Use the graph of the results in Figure 2 to determine a the resistance of the resistor and b the zero error.

Figure 2 Graph of pd against current for the resistor

For a resistor $V = IR$ where pd = V and current = I. When pd = 0, current = 0.

This graph is a straight line *not* through the origin – so it has an equation $y = mx + c$, where $y = $ pd V and $x = $ current I. So the equation is:
$$V = mI + c$$

This means m is the resistance R, which has a constant value, and c is the zero error:

$$V = RI + \text{zero error}$$

a $R = m = \text{gradient of the graph} = \dfrac{\text{change in } V}{\text{change in } I}$

On the graph, change in $V = 4.0$ and change in $I = 250$, but these are numbers on the graph, and you need to look at the units on the axes to get the quantity resistance:

$V = 4.0\,\text{V}$ $I = 250\,\text{mA} = 0.25\,\text{A}$

$\text{Resistance} = \dfrac{40\,\text{V}}{0.25\,\text{A}} = 16\,\Omega$

b Zero error $= c = $ value of V when $I = 0 = 0.4\,\text{V}$

PRACTICE QUESTIONS

Determine the gradient of the graphs to answer the following questions.

1 **B** This graph shows image size against object size.

Use the graph and the equation:

$$\text{magnification } M = \frac{\text{image height}}{\text{object height}}$$

to determine the magnification.

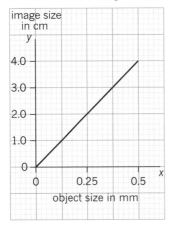

2 **P** This is a graph for heating a liquid. Use the graph to determine the rate of temperature increase in °C/s.

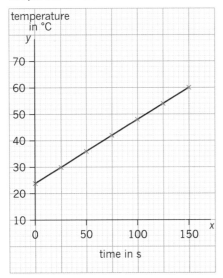

4.3 Graphs and direct proportion

Direct proportion: $y = mx$

In Topic 1.11 you saw that if one variable doubles when another variable doubles, then they are **directly proportional.**

Directly proportional: as one variable increases the other variable increases at the same rate.

For example, the extension of a spring x is directly proportional to the force F on the spring, providing it is not stretched too far:

$$F = kx$$

where k is the spring constant.

The current I through a metal wire is directly proportional to the pd V across it, providing its temperature does not change:

$$V = IR$$

where R is the resistance.

Equations like these give a straight line graph through the origin (see Topic 4.1). Whenever a graph is a straight line through the origin, you have plotted two variables that are directly proportional to each other.

WORKED EXAMPLES

1 **Figure 1 shows a graph of stretching force against extension for a spring.**

   a **Describe the relationship shown by the graph.**

   b **Calculate the spring constant for the spring in N/m.**

   c **Write an equation for the line.**

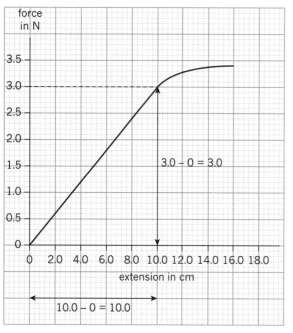

Figure 1

a The force F is directly proportional to the extension x until the force reaches 3.0 N. At higher forces, the extension increases at a faster rate as the force increases and they are no longer proportional.

b $F = kx$ so k = gradient of straight line

$$= \frac{\text{change in } F}{\text{change in } x}$$

When change in $F = 3.0 - 0 = 3.0$ N, change in $x = 10.0 - 0 = 10.0$ cm $= 0.10$ m.

$$k = \frac{3.0\,\text{N}}{0.10\,\text{m}} = 3.0\,\text{N/m}$$

The equation is of the form $y = mx$

$$F = 30x$$

2 **Figure 2 shows a graph of the length of a spring against the stretching force.**

   a **Calculate the length of the unstretched spring.**

   b **Calculate the spring constant.**

   c **A student says, 'The graph is a straight line so length is directly proportional to stretching force.' Is the student correct? Justify your answer.**

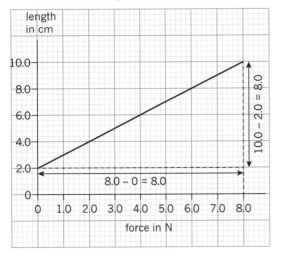

Figure 2

a When force $F = 0$, the spring will be unstretched.
From the graph: when $F = 0$, length $= 2.0\,cm$.
So unstretched length $= 2.0\,cm$

b The change in length is the same as the change in extension.
From the graph:
change in force $F = 10.0 - 2.0 = 8.0\,N$
change in length = change in extension
$$= 10.0 - 2.0 = 8.0\ cm$$
$$= 0.08\ m$$
spring constant $k = \dfrac{8.0\,N}{0.08\,m} = 100\,N/m$

c The student is wrong. The graph does not go through (0,0), so length is not directly proportional to stretching force. The relationship is **linear** but not directly proportional.

PRACTICE QUESTION

1 **P** This graph shows how the mass of some solid copper objects vary with their volume.

   a Use the graph to determine the density of copper.

   b Write an equation for the line.

   c Describe the relationship between the mass and the volume of a copper object.

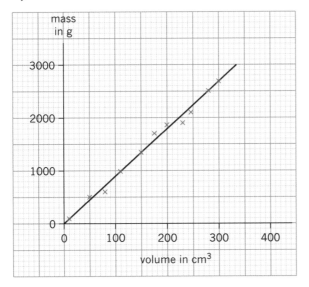

4.4 Graphs and inverse proportion

Inverse proportion

When two quantities are linked so that if one quantity doubles the other halves, they are **inversely proportional**.

Inversely proportional: one value decreases at the same rate that the other value increases.

For example:

1   The rate of photosynthesis of a plant is inversely proportional to the distance from the light source squared. This is because the intensity of the light follows an inverse square law (see Topic 1.12).

   If you double the distance between the light source and the plant, the rate of photosynthesis will decrease to one quarter of what it was. This is because $\frac{1}{2^2} = \frac{1}{4}$:

   $$\text{rate of photosynthesis} \propto \frac{1}{(\text{distance between plant and light source})^2}$$

2   The pressure of a sample of gas is inversely proportional to its volume.

   This is only true if the temperature doesn't change *and* the gas is not compressed so much that it starts turning to a liquid.

   When the gas is compressed into a volume half the size (Figure 1), the pressure is doubled:

   $$\text{pressure} \propto \frac{1}{\text{volume}}$$

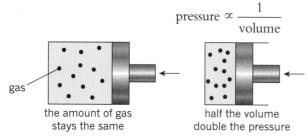

the amount of gas stays the same

half the volume double the pressure

Figure 1

WORKED EXAMPLE

A syringe was used to compress a gas slowly so that its temperature stayed constant. The volume and pressure were measured, and a graph of pressure against volume was drawn.

a   Use the graph (Figure 2) to describe the relationship between pressure and volume.

b   Draw a suitable graph to evaluate whether pressure is inversely proportional to volume.

Pressure in kPa	Volume in cm³	$\frac{1}{volume}$ in 1/cm³
100	82	0.012
120	68	0.015
140	58	0.017
160	50	0.020
180	45	0.022
200	41	0.024

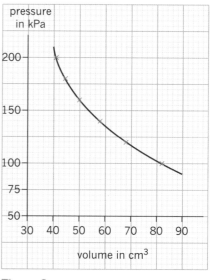

Figure 2

a The graph shows that volume decreases as pressure increases.
There is a negative correlation but it is not a linear, or directly proportional, relationship. At higher pressures, the decrease in volume is less for the same increase in pressure (or, in other words, the rate of decrease of volume is smaller).

Notice that you can't tell from the curve whether it is an inversely proportional relationship. There are too many different types of curve it could be. If you try multiplying pressure by volume, you should get the same constant number each time – but this is experimental data, so the number is close, but not exactly the same. It is better to plot a graph of pressure against $\frac{1}{\text{volume}}$ and see if it is a straight line through the origin.

That is what you are being asked to do in part **b**.

b **Step 1:** add a column to the table and calculate 1 ÷ volume for each value.

Step 2: choose your axes. Remember, you need to see if the line goes through (0,0), so you must include (0,0) on the graph.

Step 3: plot the points and draw the best straight line through the points (see Topic 2.12). Try to include (0,0) if you can, but don't force the line through that point if it doesn't fit with the others because there may be a zero error (see Topic 4.2).

Step 4: don't forget to state your conclusion: The graph of pressure against $\frac{1}{\text{volume}}$ is a straight line through the origin, so pressure is inversely proportional to volume.

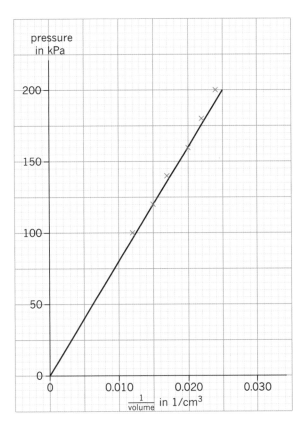

Figure 3

 PRACTICE QUESTIONS
See Topic 4.9 Further questions on graphs.

4.5 Displacement–time graphs

Displacement–time graphs

This topic looks at graphs of displacement plotted against time. You can determine velocity and acceleration from these graphs.

Distance–time graphs give speed and acceleration in the same way. See Topic 5.5 for more about vectors and scalars and the difference between speed and velocity or distance and displacement.

On displacement–time graphs:

- a horizontal line means the object is stationary, because the displacement is not changing as time changes
- a linear slope means constant velocity, because in equal times the object travels equal displacements
- a curve means acceleration. You can see this on Figure 1.

> **NOTE:** 1 small square = 50 m. So in the first 30 s, the bicycle travels 3 × 50 m = 150 m.
>
> In the second 30 s, the bicycle travels (8 × 50 m) – 150 m = 400 – 150 m = 250 m. This is greater than 150 m, so the bicycle is accelerating.

WORKED EXAMPLE

Figure 1 shows data for a bicycle journey along a straight road. Describe the motion of the bicycle in sections A, B, C, D and E.

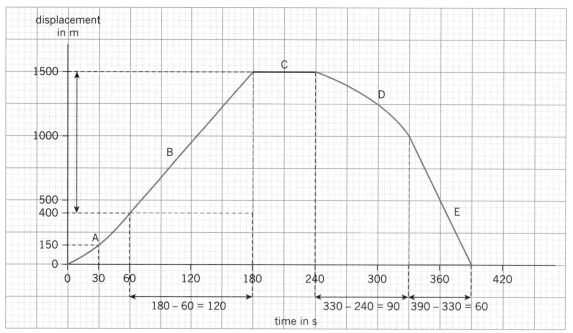

Figure 1 A bicycle journey

A: for the first 60 s, the graph curves and is getting steeper, so the bicycle is accelerating.

B: for the next 120 s, there is a straight line sloping upwards. This is constant velocity away from the start.

C: for the next 60 s, the line is horizontal, so the bicycle is stationary.

D: for the next 90 s, the displacement is getting smaller, so the bicycle is travelling back towards the starting point. The line is curving down and is getting steeper, so the bicycle is accelerating.

E: for the last 60 s, there is a straight line sloping downwards. This is constant velocity back towards the starting point.

The slope or gradient of a straight line

$$\text{average speed } v \text{ (m/s)} = \frac{\text{distance } s \text{ (m)}}{\text{time } t \text{(s)}} \qquad \text{average velocity } v \text{ (m/s)} = \frac{\text{distance } s \text{ (m)}}{\text{time } t \text{(s)}}$$

On the graph (Figure 2):

$$\text{slope} = \frac{\text{change in displacement on the vertical axis}}{\text{time taken}}$$

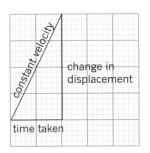

You can find out more about slopes (or gradients) of lines in Topic 4.1.

average velocity = slope of displacement–time graph

average speed = slope of distance–time graph

If the speed, or velocity, is constant, then this will be the same as the average speed, or velocity.

Figure 2

WORKED EXAMPLE

For the journey in Figure 1, calculate the velocity in a section B and b section E.

a Slope of B in the red triangle $= \dfrac{\text{change in displacement}}{\text{time taken}} = \dfrac{1500 - 400}{180 - 60} = \dfrac{11000}{120} = 9.2$

Velocity in section B = 9.2 m/s

b The slope in part E is negative. Time is increasing, but displacement is decreasing.

Slope of E in the red triangle $= \dfrac{\text{change in displacement}}{\text{time taken}} = \dfrac{0 - 1000}{390 - 330} = \dfrac{-1000}{60} = -16.7$

Velocity in section B = −16.7 m/s or 16.7 m/s back towards the start.

PRACTICE QUESTION

1 **P** This graph shows a car journey.

  a Describe the motion of the car.

  b When does the car have the fastest velocity?

  c Calculate the average velocity in km/h in the first two hours.

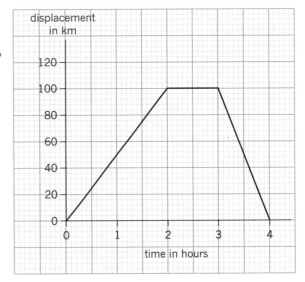

4.6 Velocity–time graphs

Velocity–time graphs

This topic looks at graphs of velocity plotted against time. Speed–time graphs can be plotted in the same way.

On velocity–time graphs:

- a horizontal line means the velocity is not changing
- a straight line sloping upwards means the object is accelerating with a constant acceleration
- a straight line sloping downwards means the object is slowing down or decelerating; the deceleration is constant
- negative velocity means the object is moving in the opposite direction
- curved lines mean the acceleration is changing.

Compare this with a displacement–time graph (see Topic 4.5). For example, a horizontal line on a displacement–time graph means the object is not moving, but on a velocity–time graph it means that the object has a constant speed.

You can see that it is very important to check carefully whether the graph is a displacement–time graph or a velocity–time graph.

WORKED EXAMPLE

Figure 1 shows a graph for a car travelling on a motorway.

a **Describe the motion of the car in sections A, B, C, D, E and F.**

b **Calculate the acceleration in i section A and ii section C.**

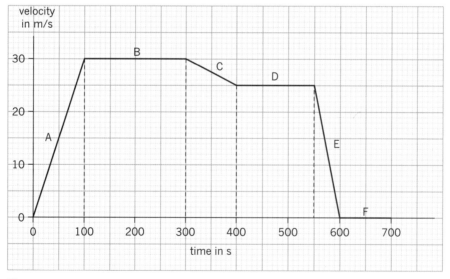

Figure 1

a **A:** the car is stationary and accelerates to 30 m/s with constant acceleration.
B: it travels at constant velocity for 300 − 100 = 200 s.
C: it slows down with a constant deceleration for 100 s.
D: it travels at constant velocity (a lower velocity than in B) for 550 − 400 = 150 s.
E: it slows to a stop with constant deceleration over 600 − 550 = 50 s.
F: it is stationary for 700 – 600 = 100 s.

b i acceleration = slope

slope of section A = $\dfrac{30 - 0}{100 - 0}$ = 0.3   acceleration in section A = $\dfrac{30 \, \text{m/s}}{100 \, \text{s}}$ = 0.3 m/s²

ii slope of section C = $\dfrac{20 - 30}{400 - 300} = \dfrac{10}{100} = -0.1$ (It is negative because this is a deceleration.)

The car is slowing down:

deceleration = $\dfrac{10 \, \text{m/s}}{100 \, \text{s}}$ = 0.1 m/s² (or acceleration = −0.1 m/s²)

PRACTICE QUESTIONS

1 **P** This is a graph showing part of a bicycle journey. Determine:

a the velocity in section A

b the acceleration in section B.

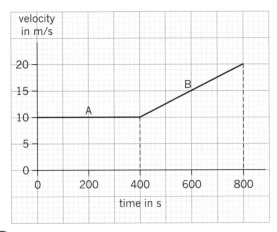

2 **P** This is a graph showing the motion of a horse.

a Describe the motion of the horse.

b Determine **i** the acceleration, **ii** the maximum velocity and **iii** the deceleration.

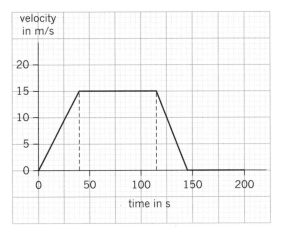

4.7 Area under a graph

The area under a line

The distance travelled by an object at constant speed is equal to the speed × time of travel. Figure 1 is a graph of speed against time for a car. On the graph, speed × time is the area of the shaded rectangle.

So the distance travelled is represented by the area of the shaded rectangle.

When an object is changing speed, the graph is not a straight horizontal line, but the distance travelled is still the area between the line and the time axis.

This method can be used for other graphs too. For example:

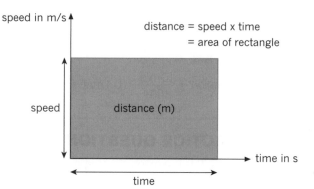

Figure 1 A graph of speed against time

- For a graph of force against extension of a spring, the area = energy stored in the spring.
- For a graph of output power of a motor against time, the area = work done by the motor.

When you calculate the area, you must use the scale on each axis. You are not working out the area of graph paper in cm^2, you are working out what that area represents.

In Figure 1, the graph is speed in metres per second against time in seconds, so the area represents distance in metres.

See Topic 5.1 for how to calculate the area of shapes.

 WORKED EXAMPLE

Figure 2 is a graph of speed against time for a bicycle. Use the graph to calculate the distance travelled a in the first 2 hours, b in the last hour and c during the whole journey.

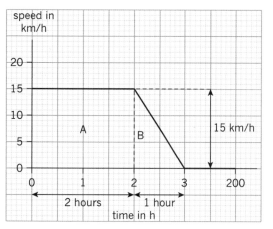

Figure 2

> **REMEMBER:** It is not the area in cm^2 you are calculating, but the area using the scale and units of each axis.

a

Step 1: write down that the distance is the area, this is worth a mark: In the first 2 hours, the distance travelled is area of rectangle A.

Step 2: read off the values on the axes:
speed = 15 km/h time = 2 hours

Step 3: calculate the distance: distance = 15 km/h × 2 hours = 30 km

b

In the last hour, distance travelled = area of triangle B, where:
area of a triangle = 0.5 × base × height
Base = time = 1 hour   height = speed = 15 km/h
Distance = area = 0.5 × 1 hour × 15 km/h = 7.5 km

c: method 1

Total distance travelled = area A + area B = 30 km + 7.5 km = 37.5 km

c: method 2

The shape is a trapezium, so you can use the equation for a trapezium. In this example, you have already calculated A and B, so method 1 is quickest. But if the question asked only for the total distance:

$$\text{area of trapezium} = \frac{(\text{side 1} + \text{side 2})}{2} \times \text{height}$$

Side 1 = time = 2 hours   side 2 = time = 3 hours   height = speed = 15 km/h

$$\text{Distance} = \text{area} = \frac{(2 + 3)}{2} \times 15 = \frac{5 \times 15}{2} = 37.5 \text{ km}$$

PRACTICE QUESTIONS

1 **P** This is a graph of speed against time for someone walking. Calculate the total distance walked.

2 **P** This is a graph of a car stopping from a speed of 30 m/s (about 70 mph). Calculate **a** the thinking distance, **b** the braking distance and **c** the total stopping distance.

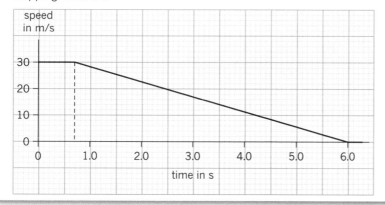

4.8 Area under a graph 2

The area under a force–extension graph

$$\text{work done} = \text{energy transferred} = \text{force} \times \text{distance}$$

When work is done to stretch something elastic, for example a spring, energy is stored in the spring and can be released when the force is removed.

The amount of energy depends on the size of the stretching force and the distance it is stretched – that is, the extension.

This means that the area between the line on a force–extension graph and the extension axis represents the energy stored in the spring.

WORKED EXAMPLE

Figure 1 is a graph of force against extension for a spring. Determine the energy stored in the spring when the extension is 3.0 cm. Give your answer in joules (J).

When the extension is 3.0 cm, the force = 2.0 N.
The energy stored is the area of the shaded triangle shown on the graph.

$$\text{area triangle} = 0.5 \times \text{base} \times \text{height}$$

Energy stored = $0.5 \times 3.0\,\text{cm} \times 2.0\,\text{N} = 3.0\,\text{Ncm}$

1 J = 1 Nm, so you must convert cm to m: 1 cm = 0.01 m

Energy stored = $3.0 \times 0.01\,\text{Nm} = 0.03\,\text{J}$

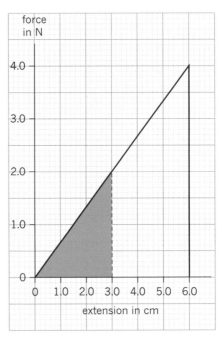

Figure 1

PRACTICE QUESTION

1 **P** Use the graph in Figure 1 to determine the energy stored in the spring when:

a the extension is 5.0 cm

b the stretching force is 2.5 N.

The area under a curve

When the line of a graph is a curve, the area can be found by counting the squares under the curve.

You work out what the area of one square on the graph paper represents and multiply by the total number of squares.

Here are some useful tips.

* It is a good idea to mark the squares as you count – so that you don't forget which ones you have included.

- You can still use the areas of rectangles and triangles to save counting large blocks of squares.
- Include all the squares where half or more of the square is under the curve.
- Leave out all the squares that have less than half a square under the curve.

WORKED EXAMPLE

Figure 2 is the graph showing a short journey by car. Calculate the distance travelled.

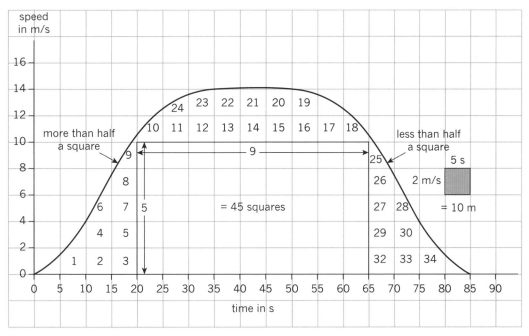

Figure 2

Step 1: use the scales to work out what the area of 1 square represents.
As shown on Figure 2: 1 square = 2 m/s × 5 s = 10 m

Step 2: mark out any large blocks that can be easily calculated. For example:

the rectangle is 9 × 5 squares = 45 squares

Step 3: count all the squares that are more than half under the curve. Numbering each one is a good way to keep count.
There are 34 squares.

Step 4: add up the areas = 45 squares + 34 squares = 79 squares

Step 5: multiply the number of squares by the area of 1 square:
distance = 79 × 10 m = 790 m

PRACTICE QUESTIONS

See Topic 4.9 Further questions on graphs.

4.9 Further questions on graphs

PRACTICE QUESTIONS

Graphs $y = mx$

1 **C** This is a graph of the volume of hydrogen gas produced during the first 30 s of a reaction. Use the graph to determine the reaction rate for the first 20 s. Give your answer in dm^3/s.

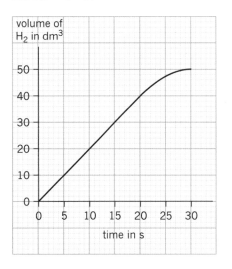

Inverse proportion

2 **B** Some students measure the rate of photosynthesis with a point light source at different distances from the plant. These are their results.

Distance in m	Rate of photosynthesis in arbitrary units
0.20	20.0
0.40	5.0
0.60	2.0
0.80	1.0
1.00	0.8

a Draw a graph of rate of photosynthesis against $\dfrac{1}{distance^2}$. Draw a line of best fit.

b Comment on what your graph shows about the relationship between rate of photosynthesis and distance from the light source.

3 **P** Some students carry out an experiment to measure how the count rate from a radioactive source changes with time. They think that the count rate is inversely proportional to the time.

a Draw a graph of $\dfrac{1}{count\ rate}$ against time. Draw a best fit line.

b Judge whether the students are correct. (For more about radioactive decay, see Topic 4.12.)

Time in minutes	Count rate in counts/s	$\dfrac{1}{\text{count rate}}$ in 1/counts per second
10	4.40	0.23
20	2.20	0.45
30	1.10	0.91
40	0.55	1.82
50	0.28	3.64

Displacement–time graphs

4   **P** This graph shows the journey made by a lorry.

    a What is the displacement of the lorry at the end of the journey?

    b Describe the motion of the lorry in sections A, B, C and D.

    c Calculate the velocity of the lorry in section B. Give your answer in km/h.

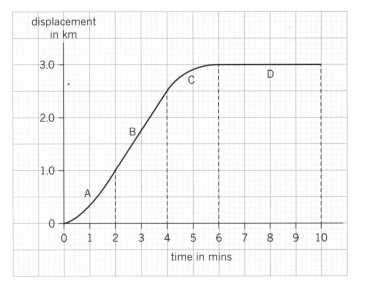

Area under a curve

5   **P** This is the speed–time graph of a short motorway journey by car. Determine the total distance travelled.

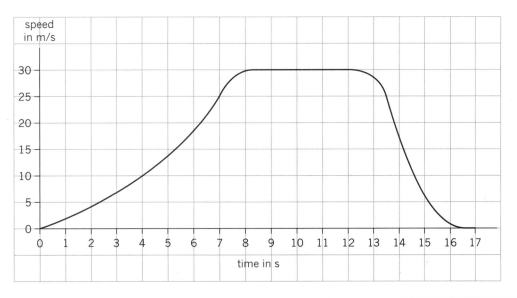

4.10 Tangents

The slope of a graph

Figure 1 shows a sketch graph of the product formed in a reaction plotted against time since the reaction began.

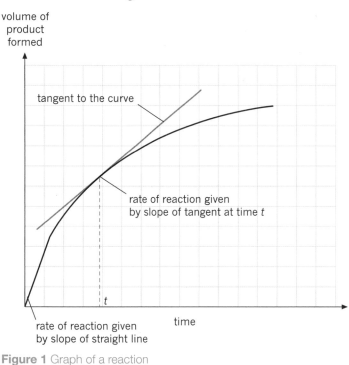

Figure 1 Graph of a reaction

In question 1 in Topic 4.9, you used the slope of the straight line at the start of the reaction to determine the rate of reaction.

$$\text{rate of reaction} = \frac{\text{amount of product formed}}{\text{time taken}} = \text{slope of graph}$$

In Figure 1, you can see that the slope changes as the graph is a curve. If you want to know the rate of reaction when the graph is curved, you need to determine the slope of the curve.

The tangent is the straight line that just touches the curve. The slope of the tangent is the slope of the curve at the point where it touches the curve.

To find out the slope of the curve, you must draw a tangent at the correct point and determine the slope of the tangent.

WORKED EXAMPLE

Figure 2 is a graph of the amount of carbon dioxide formed in a reaction, plotted against time since the reaction began.

Determine the rate of reaction 40 minutes after the reaction began.

Step 1: find the point on the curve where you need to draw the tangent.

Step 2: use a transparent ruler, as it makes it much easier when you can see the graph under the ruler.

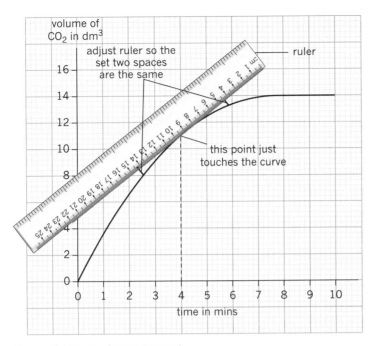

Figure 2 How to draw a tangent

Step 3: line up the ruler so that the space between the ruler and the curve is the same on each side of the curve. Turn the page around if that helps you to see better.

Step 4: use a sharp pencil to draw the line. Make it cover as much of the graph paper as you can.

Step 5: determine the slope of the straight line (see Topic 4.1).

$$\text{Slope} = \frac{16 - 4.8}{7.2 - 0} = \frac{11.2}{7.2} \qquad \text{Reaction rate} = \frac{11.2\,dm^3}{7.2\,min} = 1.6\,dm^3/min$$

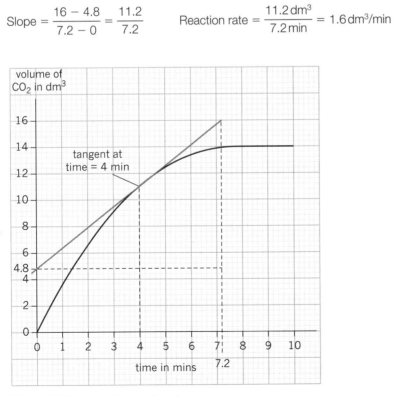

Figure 3 How to calculate the slope

4.11 Graphs of waves

Transverse waves

A transverse wave, for example a water wave, can be drawn on graph paper so that measurements of wavelength and amplitude can be made (Figure 1).

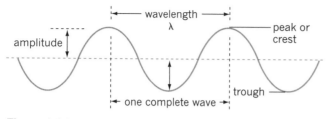

Figure 1 A transverse wave

WORKED EXAMPLE

Figure 2 is a graph showing a water wave. Determine:

a   the wavelength of the wave   **b**   the amplitude of the wave.

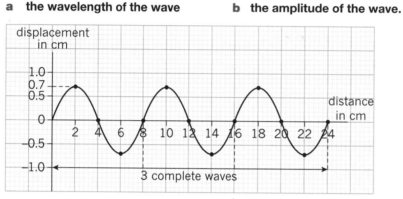

Figure 2

a   The wavelength is one complete wave. You can measure this from any point on the wave to the next point where it repeats, but it is easiest to do crest to crest, trough to trough, or where the wave crosses the axis (but take care not to do half a wave if you choose this point).

If there are lots of waves, using several will give a more accurate answer:

3 complete waves = 24 cm, so 1 wavelength = $\dfrac{24\,\text{cm}}{3}$ = 8 cm

b   Amplitude = 0.7 cm

PRACTICE QUESTION

1   Ⓟ Determine **a** the wavelength and **b** the amplitude of this wave.

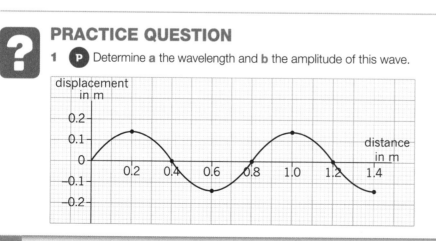

Displacement–time graphs

You can plot graphs of the displacement (how much the wave goes up and down, or oscillates) against time.

When you look quickly at these graphs, they look almost the same as the graphs of displacement against distance. It is important to look carefully at the horizontal axis to see which type of graph it is.

You can determine the amplitude from these graphs, but not the wavelength.

The time for one complete oscillation is called the period, and this is measured from any point on the graph to where it repeats. This is also called one cycle.

The period is measured in seconds.

You can use this to calculate the number of waves in a second. This is the frequency:

$$\text{frequency} = \frac{1}{\text{period}}$$

It is measured in Hz.

If one complete wave takes half a second (0.5 second), then there will be two waves in a second (2 per second = 2 Hz).

WORKED EXAMPLE

Figure 3 is a graph showing a wave. Determine:

a   **the amplitude**

b   **the period**

c   **the frequency of the wave.**

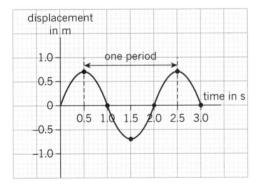

Figure 3

a   The amplitude = 0.7 cm

b   The period = one complete cycle; peak to peak is the easiest to measure
  = 2.5 s − 0.5 s = 2.0 s

c   Frequency = $\dfrac{1}{\text{period}} = \dfrac{1}{2.0} = 0.5\,\text{Hz}$

4.12 Half-life

Radioactive decay

You cannot tell when a radioactive nucleus will decay. It might be within the next second, or it might still be here in a million years. However:

- if you take a sample of a million carbon-14 nuclei, then, on average, there will be only half a million left in 5740 years' time

- if you take a sample of a million oxygen-15 nuclei, then, on average, there will be only half a million left in 2 minutes' time.

This time is called the half-life of the radioactive isotope.

It is different for each radioactive isotope.

You can calculate how many nuclei will be left after a certain time if you know the half-life. And you can calculate the half-life if you know how many nuclei are left after a certain time.

WORKED EXAMPLE

Technetium-99m has a half-life of 6.0 hours. There are 3 840 000 nuclei in a sample. Calculate how many are left after 24 hours.

Method 1

Step 1: calculate the number of half-lives that have passed:

$$\text{number of half-lives} = \frac{\text{total time}}{1\ \text{half-life}} = \frac{24\ \text{hours}}{6.0\ \text{hours}} = 4$$

Step 2: after 1 half-life, the number will have halved: $\frac{3\,840\,000}{2} = 1\,920\,000$

Step 3: repeat this step three more times, so that you have halved the number 4 times. This is easy to do if you have a calculator and the number of half-lives is small.

$$\frac{1\,920\,000}{2} = 960\,000 \text{ nuclei after three half-lives}$$

$$\frac{960\,000}{2} = 480\,000 \text{ nuclei after two half-lives}$$

$$\frac{480\,000}{2} = 240\,000 \text{ nuclei after four half-lives}$$

Method 2

Step 1: calculate the number of half-lives that have passed:

$$\text{number of half-lives} = \frac{\text{total time}}{1\ \text{half-life}} = \frac{24\ \text{hours}}{6.0\ \text{hours}} = 4$$

Step 2: after 4 half-lives, the number will have halved 4 times, so divide by $(2 \times 2 \times 2 \times 2) = 2^4$

$$\frac{3\,840\,000}{2^4} = \frac{3\,840\,000}{16} = 240\,000 \text{ nuclei}$$

WORKED EXAMPLE

A sample contains 100 000 iodine-131 nuclei. After 24 days, there are 12 500 left. Calculate the half-life of iodine-131.

Count the number of times you have to halve 100 000 to get 12 500.

$1: \dfrac{100\,000}{2} = 50\,000$

$2: \dfrac{50\,000}{2} = 25\,000$

$3: \dfrac{25\,000}{2} = 12\,500$

answer = 3 half-lives

24 days = 3 half-lives, so 1 half-life $= \dfrac{24}{3} = 8$ days

PRACTICE QUESTIONS

1 (P) A sample contains 6 million cobalt-60 nuclei. After 21 years, there are 375 000 left. Calculate the half-life.

2 (P) Nuclei of the gas radon-222 have a half-life of 3.8 days. Calculate how many nuclei of a sample of 2.56×10^6 nuclei are left after 26.6 days.

4.13 Graphs of half-life

Radioactive decay curves

When you plot the number of radioactive nuclei in a sample against time you get a curve showing that the number decreases, because some nuclei have decayed.

The activity of a radioactive sample is the number of nuclei that decay per second. This is the rate at which the radiation is emitted and it is measured in counts per minute or counts per second.

As time passes, the activity of a radioactive sample decreases, because the number of radioactive nuclei left to decay is decreasing. When you plot the activity of the sample against time you get exactly the same shape decay curve as the decay curve showing the number of nuclei that are left.

Half-life

The time for the activity of the sample to decrease to a half of its original value, or for the number of nuclei to halve, is the half-life.

A curve like this, where the number always halves in the same time interval, is called an exponential curve. (Figure 1).

WORKED EXAMPLE

Use Figure 1, a graph of activity against time, to determine the half-life of sodium-24.

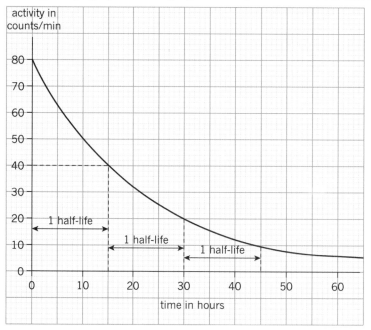

Figure 1

Method 1

Step 1: choose a starting point, for example, the value when time = 0 the activity
= 80 counts/min.

Step 2: activity ÷ 2 = 80 ÷ 2 = 40 counts/min.

Step 3: find the time when the activity has this value (40 counts/min): time = 15 h

Step 4: half-life = time 2 – time 1 = 15 – 0 = 15 h

Method 2

To reduce the effect of random differences, find the average half-life by finding the time for more than one half-life:

Step 1: choose a starting point, for example, the value when time = 0 the activity = 80 counts/min.

Step 2: choose a number of half-lives, for example 3, and work out the activity:

$80 \div 2^3 = 80 \div 8 = 10$ counts/min

Step 3: find the time when the activity has this value: time = 45 h

Step 4: 3 half-lives = time 2 – time 1 = 45 – 0 = 45 h

Step 5: 1 half-life = 45 ÷ 3 = 15 h

PRACTICE QUESTION

1 **P** This graph shows the number of nuclei remaining plotted against time for a radioactive isotope. Use the graph to determine the half-life.

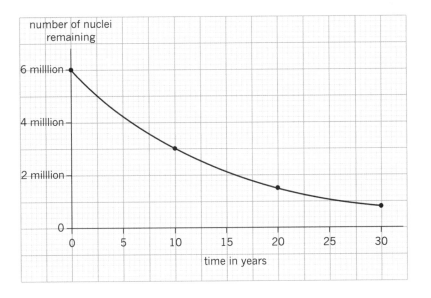

5 GEOMETRY

5.1 Area

Calculating area

Lengths are measured with a ruler. Areas are measured with squares – one centimetre squares ($1\,cm^2$) or 1 metre squares ($1\,m^2$). But you don't carry a square with you. Instead, you calculate the number of squares.

Learn the equations for calculating areas (see Topic 6.5).

WORKED EXAMPLE

The squares marked on all of the shapes in Figure 1 are $1\,cm^2$ (not to scale).

Calculate the areas of a to d and estimate the area of e.

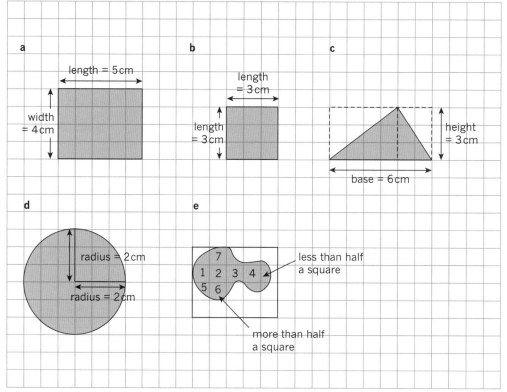

Figure 1

a The area of the rectangle = length × width
    $$= 5 \times 4 = 20\,cm^2$$

b The area of the square = length^2
    $$= 3 \times 3 = 9\,cm^2$$

c The area of the triangle = 0.5 × base × height
    $$= 0.5 \times 6 \times 3 = 9\,cm^2$$

d The area of the circle = π × (radius)^2
    $$= \pi \times 2^2 = 12.6\,cm^2$$

e The area of the irregular shape can be estimated by counting squares. Include squares that are more than half a square and ignore squares that are less than half a square. Number them to help you keep count.
    Area = ~$7\,cm^2$

> **NOTE:** Notice that the shaded triangle is half the area of the rectangle, with area base × height.

Surface areas

A cube has six faces – that is why dice are numbered 1 to 6, because there is one number on each face.

To calculate the surface area of a solid shape, you must add up the area of all the faces.

 WORKED EXAMPLE

The squares marked on all of the shapes in Figure 2 are 1 cm² (not to scale).

Calculate the surface areas of **a** and **b**.

a b

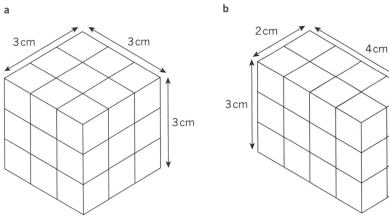

Figure 2

a The area of one face of the cube = 3 × 3 = 9 cm²

There are 6 faces. The total surface area = 6 × 9 = 54 cm²

b The top and bottom faces are 2 × 4 = 8 cm²

The two largest faces are 4 × 3 = 12 cm²

The two smallest faces are 2 × 3 = 6 cm²

The total 6 faces = (2 × 8) + (2 × 12) + (2 × 6) = 16 + 24 + 12 = 52 cm²

 PRACTICE QUESTIONS

1 **B** Estimate the area of a leaf by treating it as a triangle with base 2 cm and height 9 cm.

2 **B** Estimate the area of a cell by treating it as a circle with a diameter of 0.7 µm. Give your answer in µm².

3 **C** Calculate the surface area of cubic marble chips with sides
a 0.5 cm   **b** 0.8 cm.

4 **P** Use the equation:

$$\text{pressure} = \frac{\text{force}}{\text{area}}$$

to calculate the pressure exerted by a board when a 16 N weight is placed on it and the board **a** has an area of 16 cm², **b** is a square with sides 5 cm and **c** is a triangle with base 10 cm and height 8 cm.

Give your answers in N/cm² and in Pa.

For more practice calculating areas, see Topics 4.7 and 4.8.

> **REMEMBER:** Check carefully if you have been given an area or a length. If you have been given the area of a square, do not square the value again.

5.2 Volume and surface area to volume ratios

Calculating volume

Volumes are measured in one centimetre cubes ($1\,cm^3$), one decimetre cubes ($1\,dm^3$) or one metre cubes ($1\,m^3$). To find the volume for some shapes, you can calculate the number of cubes.

✓ WORKED EXAMPLE

The cubes marked on all of the shapes in Figure 1 are $1\,cm^3$ (not to scale).

Calculate the volumes of a and b.

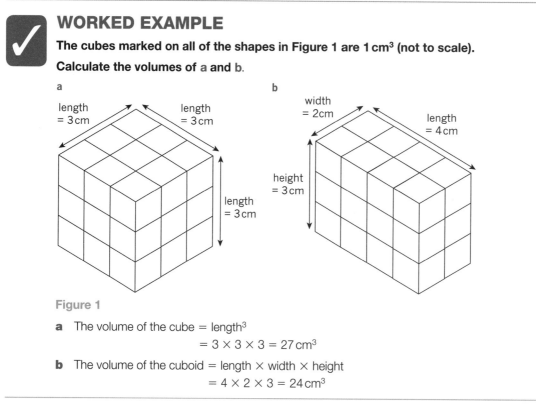

Figure 1

a The volume of the cube = length³
$$= 3 \times 3 \times 3 = 27\,cm^3$$

b The volume of the cuboid = length × width × height
$$= 4 \times 2 \times 3 = 24\,cm^3$$

Cross-sectional areas

When a slice is cut straight through a solid shape, the area of the slice is called the cross-sectional area. Some examples of these are shown in Figure 2.

The volume of shapes like these = cross sectional area × length

---

✓ WORKED EXAMPLE

Calculate the volumes of the shapes in Figure 2.

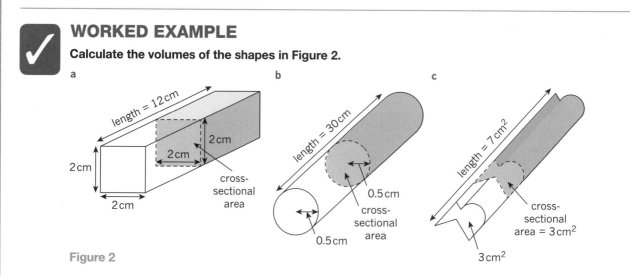

Figure 2

The volume of each shape = cross-sectional area × length

a Volume = area of square × length
= 2 × 2 × 12 = 48 cm³

b Volume = area of circle × length
= π × 0.5 × 0.5 × 30 = 23.6 cm³

c Volume = area of cross-section × length
= 3 × 7 = 21 cm³

Surface area to volume ratio

The surface area to volume ratio is written as follows.

$$\text{surface area : volume} \qquad \text{or} \qquad \frac{\text{surface area}}{\text{volume}}$$

The units are m²/m³ or you can use cm²/cm³. If you are comparing different shapes, use the same units.

WORKED EXAMPLE

Calculate the surface area to volume ratio for a cube with sides of length 4 cm.

Surface area = 6 × 5 × 5 = 150 cm² (see Topic 5.1)

Volume = length³ = 5 × 5 × 5 = 125

Surface area to volume ratio = $\dfrac{150}{125}$ = 1.2 cm²/cm³ or 1.2 : 1

PRACTICE QUESTIONS

1 **B** Estimate the volume of a red blood cell by assuming it is a cube of length 0.7 μm in μm³.

2 **C** Calculate the surface area to volume ratio for the following marble chips.

   a cubes of length 0.1 cm

   b cubes of length 10 cm

3 **B** Estimate, by assuming they are cubes, the surface area to volume ratios of:

   a an amoeba, length 2.50 × 10⁻⁴ m (state your answer in standard form)

   b a human, length 2 m

   c a whale, length 25 m.

4 **P** Use the equation:

$$\text{mass} = \text{density} \times \text{volume}$$

   to calculate the mass of:

   a a cube of concrete with sides of length 0.5 m
   (density of concrete = 2400 kg/m³)

   b a steel rod 2 m long with square cross-section of sides 1 mm
   (density of steel = 8000 kg/m³)

   c a copper wire 5 m long with circular cross-sectional area 0.1 mm²
   (density of copper = 8900 kg/m³).

5.3 Angles

Angles are measured in degrees. The symbol is a small circle. For example, 60 degrees = 60°.

Angles are measured with a protractor. This is not listed as required for most written exams, but you do need to be able to determine the size of some angles.

Parallel = 0°

When there is no angle between two lines, they are in the same direction.

A right-angle = 90°

This is the angle between a vertical line (down towards the centre of the Earth) and a horizontal line (a level line that is not sloping).

Two lines at an angle of 90° are also called **perpendicular lines**. Sometimes, for example in ray diagrams, the line at a right-angle to a surface is called a **normal**.

A straight line = 180°

When two lines meet in a straight line, then the angle between them is 180°. This is two right-angles = 2 × 90° = 180°.

A full circle = 360°

This is four right angles = 4 × 90° = 360° (Figure 1).

The angles of a triangle add up to 180°

This can help you to work out unknown angles.

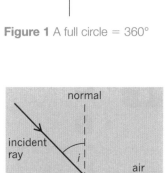

Figure 1 A full circle = 360°

Ray diagrams

Reflection

The angle of incidence i and the angle of reflection r are both angles between the normal and the ray:

angle of incidence = angle of reflection

Refraction

The angle of incidence i and the angle of refraction r are both angles between the normal and the ray.

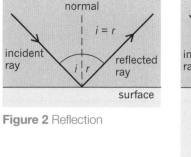

Figure 2 Reflection

Figure 3 Refraction

WORKED EXAMPLE

1 Calculate angles a to d in Figure 4 using your knowledge of angles.

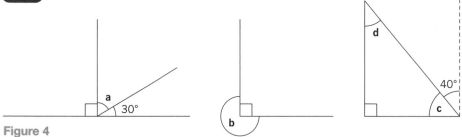

Figure 4

The angle **a** + 30° = 90°

a = 90° − 30° = 60°

The angle **b** = 360° − 90° = 270°

The angle **c** = 90° − 40° = 50°

The angles of the triangle add up to 180°:

90° + **c** + **d** = 180° and **c** = 50°

d = 180° − 90° − 50° = 40°

2 The ray diagram in Figure 5 shows a light ray being reflected. Determine **a** the angle of incidence and **b** the angle of reflection.

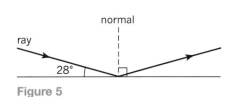

Figure 5

a The angle of incidence = 90° − 28° = 62°

b The angle of reflection = angle of incidence = 62°

3 The ray diagram in Figure 6 shows a light ray being refracted in glass. Determine **a** the angle of incidence and **b** the angle of refraction.

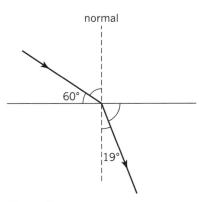

Figure 6

a The angle of incidence = 90° − 60° = 30°

b The angle of refraction = 19°

PRACTICE QUESTION

1 **P** Determine the angles **a** to **e** in the following diagrams.

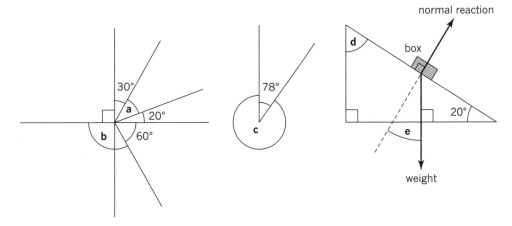

5.4 Pie charts

Using pie charts

On a pie chart, data is represented by dividing a circle into sectors. The size of each sector is proportional to the fraction or percentage of the whole 'pie'.

To draw a pie chart, you use a protractor to measure the angle for each sector.

The whole 'pie' is 360°, so 50% = 180° and 10% = 36°.

To interpret a pie chart, you usually do not need to measure the angles because the sectors are labelled, or you can tell by the shape. In Figure 1, for example, the sectors are 50%, 25%, 15% and (25 − 15) = 10%.

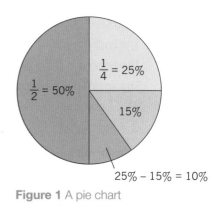

Figure 1 A pie chart

WORKED EXAMPLE

Some students estimate the percentage of different food groups in their diet. They draw a pie chart. Determine the percentage of the diet that is each of the following food groups.

a carbohydrate

b fat

c protein

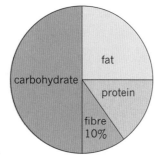

Figure 2 Balanced diet

a Carbohydrate = $\frac{1}{2}$ of the circle $\overset{\bullet}{=}$ 50%

b Fat = $\frac{1}{4}$ of the circle = 25%

c Protein = remaining 25% − 10% fibre = 15%

PRACTICE QUESTIONS

1 C This pie chart shows the sources of carbon dioxide emissions. Determine the percentage of carbon dioxide emissions from:

a domestic transport

b industrial

c residential.

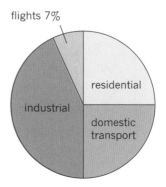

2 **C** These two pie charts represent the composition of the atmosphere of the Earth and Mars. The data is also given in the table.

Use the pie charts and the table to determine the correct answer.

a Which planet is Mars?

b Which sectors (A–I) could be the percentage of:

i nitrogen?　ii oxygen?　iii carbon dioxide?　iv argon?

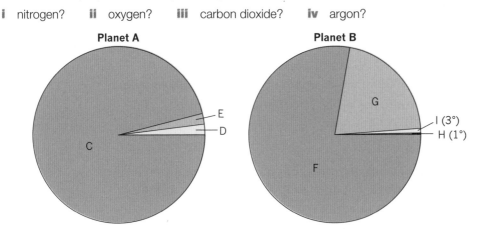

Gas	% on Earth	% on Mars
argon	0.9	2
carbon dioxide	< 0.1	96
oxygen	21.0	0
nitrogen	78.0	2

3 **B** Some students estimate the number of different species of plants and animals in their local area. Their results are shown in the diagram. Determine:

a the percentage of insects

b the total percentage of invertebrates

c the percentage of plants.

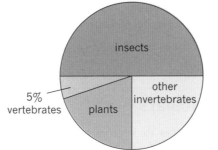

4 **P** The diagram shows the exposure to background radiation for a person in part of the UK. Determine the percentage of radiation from:

a radon gas in the air

b the total from food and drink and medical procedures

c cosmic rays.

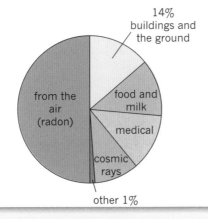

5.5 Vectors and scalars

Vectors and scalars in motion

Velocity is a **vector**. Vectors have a magnitude (size) and a direction. For example, 30 mph east and 50 km/h north-west are velocities.

Speed is a **scalar**. Scalars have a magnitude, but no direction. For example, 10 m/s is a speed.

Directions can be given as points of the compass, angles or words, such as forwards, left or right.

Distances, for example 4 m or 100 km, are scalars, because they have only magnitude.

Displacements, for example '4 m to the left' and '100 km south' are vectors – they have magnitude and direction.

Positive and negative

Sometimes velocities in opposite directions are called positive and negative velocities. Speeds should really be just positive – but negative speeds are sometimes used.

Acceleration

Acceleration is defined as rate of change of velocity (not speed), so acceleration is a vector. An object moving at constant speed in a circle is accelerating, because its direction is changing. Its velocity is changing although its speed is constant.

> **REMEMBER:** Scalars have magnitude only. Vectors have magnitude and direction.

WORKED EXAMPLE

Sort these quantities and values into vectors and scalars: acceleration, displacement, length, velocity, 50 km/h, 100 m east.

Vectors: acceleration, displacement, velocity, 100 m east

Scalars: length, 50 km/h

More vectors and scalars

A force has a direction, so all forces are vectors, including air resistance (which always has a direction opposite to the movement of the object). Momentum also has direction and is a vector.

Scalars include mass, area, time, energy, work, temperature and electric current.

You may think temperature goes up and down, but these are words used to describe increasing and decreasing – the temperature is not to the north, nor is it to the right.

Electric current is a flow of charge, so the charge is moving, but current is a scalar quantity. Current is only the magnitude of the flow of charge, it doesn't include the direction. (Remember, speed has magnitude only, but the object is moving in some direction and speed doesn't include the direction.)

PRACTICE QUESTIONS

1 (P) Sort the following quantities into a table of scalars and vectors.
   speed, velocity, force, distance, temperature, displacement, acceleration, energy, friction, mass, weight, volume, area, momentum

2 (P) Sort the following values into a list of vectors and a list of scalars.
50 km/h, 100 m east, 9 m/s, 25 m/s, 16 m, 2000 km west, 3×10^8 m/s
upwards, 273 °C, 50 kg, 30 N downwards, 3 A

How to draw vectors

Vectors are shown on drawings by a straight arrow. The arrow starts from the point
where the vector is acting and shows its direction. The length of the vector represents
the magnitude.

When 'to the right' is the positive direction:

force: ⎯⎯⎯⎯⎯ 5 N ⎯⎯⎯⎯⎯→   velocity: ←⎯ −3 m/s ⎯⎯

Adding vectors

When you add vectors, for example two velocities or three forces, you must take the
direction into account.

The combined effect of the vectors is called the **resultant**.

If the vectors are in the same or the opposite directions, the magnitudes can be added
and subtracted. Vectors can also be combined by using scale drawing (see Topic 5.7).

✓ **WORKED EXAMPLE**

Determine the resultant of the following vectors.

a   2 N →   5 N →   3 N →   =   resultant = 10 N →

b   1 N →   2 N →   −5 N ←   =   resultant = −2 N ←

? **PRACTICE QUESTIONS**

3 (P) Determine the resultant of the following forces.

a   10 N upwards and 6 N upwards

b   4 N upwards and 3 N downwards

c   1 N north and 1 N south

d   10 N →   17 N →   7 N ←   **e**   ← 3 N   ← 3 N   ← 6 N

4 (P) Determine the resultant momentum when the following objects collide and stick together.

a   6 kg m/s →   1 kg m/s →   **b**   100 kg m/s →   −50 kg m/s ←

↑ **STRETCH YOURSELF!**

5 (P) These objects collide and move off separately. Use the equation:

total momentum before collision = total momentum after collision

to calculate the missing value of momentum.

Before: 5 kg m/s →   −9 kg m/s ←   After: −2 kg m/s ←   ?

5.6 Free body force diagrams

When you are adding vectors that are not parallel, the direction is important. A free body force diagram shows all the forces and their directions and is the first step to determining the resultant force.

In a free body force diagram, the object is shown as a dot and the forces are shown as arrows that start on the dot and are drawn in the direction of the force. They don't have to be to scale, but it helps if the larger forces are shown to be larger.

WORKED EXAMPLES

1 **Figure 1 shows a falling ball. Draw a free body force diagram for the ball a accelerating downwards and b falling at terminal velocity (a constant velocity).**

a accelerating downwards

air resistance

weight

weight > air resistance

Figure 1 A falling ball

b at a constant velocity

air resistance

weight

weight = air resistance

2 **Figure 2 shows a brick on a ramp. It is not moving. Draw a free body force diagram for the brick.**

reaction

friction

weight

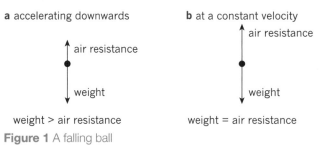

reaction   brick

friction

weight

Figure 2 A brick on a ramp

There are three forces on the brick. The reaction force is at 90° to the ramp. The friction stops the brick sliding down, so it is parallel to the ramp and up the ramp. The weight of the brick acts vertically downwards.

PRACTICE QUESTION

1 (P) Draw free body force diagrams for **a** the aeroplane in level flight at steady speed, and **b** the lamp.

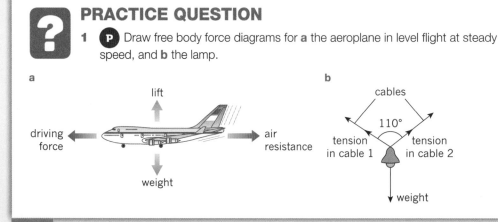

a

lift

driving force

air resistance

weight

b

cables

110°

tension in cable 1

tension in cable 2

weight

Triangles of forces

The resultant of two forces can be determined by considering one force acting after the other. The result is the same as if they both act together. Figure 3 shows how to do this.

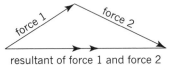

Figure 3 The resultant of two forces

Step 1: draw an arrow representing force 1. This must be in the direction force 1 is acting.

Step 2: starting at the tip of the force 1 arrow, now draw an arrow representing force 2 in the direction force 2 is acting.

Step 3: the resultant starts at the same starting point as force 1, and finishes at the tip of the force 2 arrow.

You have drawn a triangle of forces. The resultant can be calculated, or found from a scale drawing (see Topic 5.7).

It doesn't matter whether you start with force 1 or Force 2. The resultant is the same.

When there are three balanced forces on an object (for example, the brick in worked example 2), the three forces form a triangle of forces – there is no resultant force acting.

✔ WORKED EXAMPLES

3 **Two tractors are pulling a log across a field. Draw a triangle of forces to show the resultant force from the two tractors.**

    Step 1: draw force 1.

    Step 2: draw force 2 at 90° to force 1.

    Step 3: draw the resultant.

4 **In worked example 2, the forces on the brick are balanced. Draw a triangle of forces for the brick.**

    For balanced forces, draw the third force from the tip of the second back to the tail of the first.

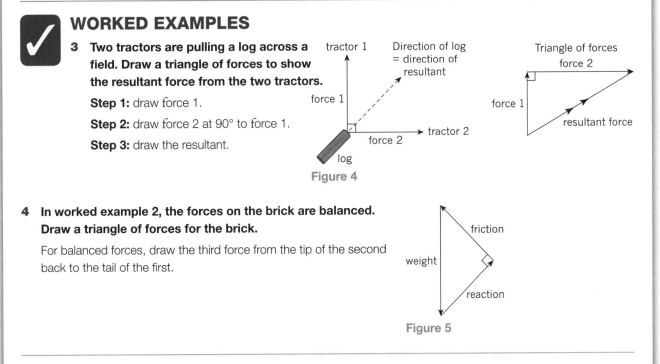

Figure 4

Figure 5

? PRACTICE QUESTION

2 Ⓟ Draw triangles of forces to show balanced forces on the objects in practice question **1b** and **c**.

5.7 Scale drawings

Vector triangles

One way to find the size and direction of the resultant force is to draw a scale diagram.

You will need a sharp pencil, a transparent ruler and, if squared paper is not provided, a protractor for measuring angles.

On exam papers, you may be given a grid of 1 cm squares and be asked to find the resultant of two forces at right angles.

 WORKED EXAMPLE

Two tug boats are pulling a ship, as shown in Figure 1. Draw a scale diagram using the squared paper provided to determine the size of the resultant force.

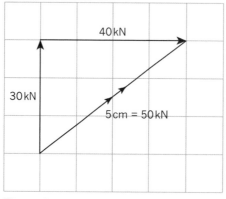

Figure 1

Step 1: choose a scale that will fit on the paper and use more than half of it.

Scale: 1 cm = 10 kN

Step 2: with your ruler, draw one of the forces to scale on the diagram.

30 kN = 3 cm

Step 3: from the tip of the arrow, draw the second force to scale on the diagram (use your ruler).

40 kN = 4 cm

Step 4: draw a straight line from the start point to the tip of the second force arrow and label this with two arrowheads to show it is the resultant.

Step 5: measure the length of the resultant with your ruler.

Length = 5 cm

Step 6: multiply the length in cm by 10 kN to get the size of the force.

Size of resultant force = 50 kN

If you do not have squared paper, then you must draw the arrows the correct length by using your ruler, and use the protractor to make sure the two forces are at 90° to each other.

Figure 2

Balanced forces

When three forces are balanced and two are known, a scale diagram can be used to find the third force.

WORKED EXAMPLE

A shop sign against the wall of a shop is supported by a cable attached to the wall, as shown in Figure 3. Draw a scale diagram to determine the reaction force from the wall. The tension in the cable has been drawn for you.

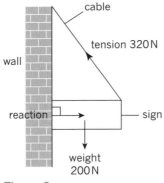

Figure 3

The tension force arrow is already drawn. Measure the arrow: length = 6.4 cm

Scale: 6.4 cm = 320 N, so 1 cm = $\dfrac{320}{6.4}$ = 50 N per cm

Draw weight arrow vertically downwards (Figure 4). Length (for 200 N): $\dfrac{200}{50}$ = 4 cm

Draw reaction arrow horizontally to the left.

Measure length of reaction arrow = 5 cm

Size of reaction force = 5 cm × 50 N per cm = 250 N

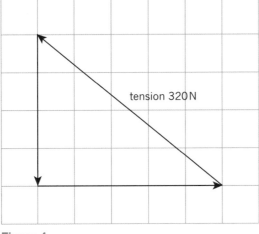

Figure 4

PRACTICE QUESTION

1 (P) In Topic 5.6 Figure 4, tractor 1 is pulling with force 1 = 5 kN and tractor 2 is pulling with force 2 = 12 kN. By scale drawing, determine the resultant force.

1 Calculate the amount of charge that flows through a resistor when a current of 450 mA passes through it for 30 minutes.

2 In a photograph, a red blood cell is 14.5 mm in diameter. The magnification stated on the image is ×2000. Calculate the real diameter of the red blood cell.

3 The graph below shows the concentration of a reactant plotted against the time from when the reaction starts.

Determine the rate of reaction in mol/dm^3 s.

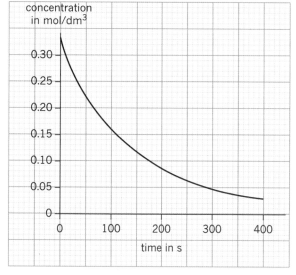

4 For the wave shown in the graph below, determine:

   a the amplitude

   b the period

   c the frequency

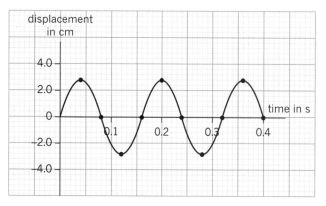

5 Calculate the area of a rectangular field 75 m long and 56 m wide.

6 Calculate the volume of water in a full swimming pool 25 m long, 12 m wide, and 2 m deep.

7 The resultant force on a ball falling downwards through the air = 3 N. The ball weighs 5 N. Calculate the force due to air resistance.

8 Use the following graph to determine the half-life of sodium-22.

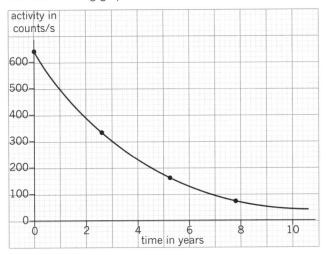

9 Use the following graph to determine the half-life of fluorine-18.

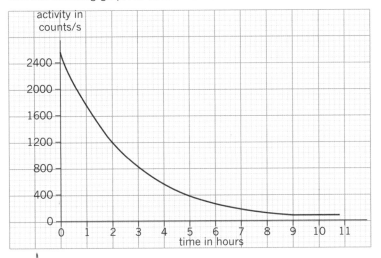

10 Determine the angles of incidence and reflection in the following ray diagrams.

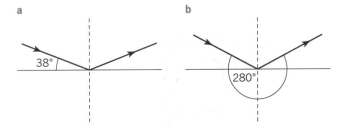

11 a Draw a free body force diagram for the climber shown in the diagram on the right.

b The climber's weight is 800 N and the reaction force is 460 N. Use a scale drawing to determine the tension in the rope.

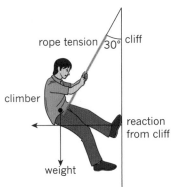

6.1 Using a calculator – the basics

Your scientific calculator

You are allowed to use a scientific calculator in the exams. It will save you time and help you to avoid mistakes in calculation – if, of course, you use it correctly.

There are different models available, but they all have similar functions. You must know in advance how to use your calculator for all the calculations you need in the exam. This will help to avoid making mistakes and wasting time.

Here are some important tips:

Tip 1: get a calculator and use it in lessons and for homework. This will make you an expert, able to use it quickly and accurately.

Tip 2: don't forget it on the day of the exam.

Tip 3: write the calculation you are going to do on the paper, and then key it carefully into the calculator. This will ensure that you get the marks for your working. You will also be less likely to make a mistake entering the calculation.

Tip 4: read the answer carefully, write it down and then check the display again. Have you written it correctly? For example, sometimes students write a 3 instead of an 8 or miss out a digit.

The 'User Guide'

Keep the user guide for reference. If you have lost it, try an internet search for the model number, as lots of user guides are available online. You could download the guide and keep it on your phone or laptop. Most people don't want to work through the user guide from cover to cover, but you may need it to refer to when you first start new types of calculation.

This guide has been written using a very common type of scientific calculator. Some of the buttons might be different on your calculator so use these hints and your user guide to find the exact keys to press.

The keypad

A scientific calculator has the same keys as a basic calculator – and then a lot of extra ones with different functions marked on them, and more functions written above the keys (Figure 1). To use the functions above the keys, you must press the shift key `SHIFT` (at the top left) before pressing the key you want. On some models you must press `SHIFT` again to switch it off.

You don't have to know how to use all the functions – you can just use the calculator like any basic calculator – but some of them are very useful.

The replay key `REPLAY` is used to move around the display. Some calculators have ▶ and ◀ instead, and these will be used to show the part of the replay key to press in this guide.

Figure 1 A scientific calculator

Getting started

Start by doing a few calculations that you have done on an ordinary calculator, and for which you know the answers. This helps to see how things are displayed.

Try $\boxed{+}$ $\boxed{-}$ $\boxed{\times}$ $\boxed{\div}$ and $\boxed{=}$

Useful keys

$\boxed{AC}$ stands for All Clear, and it deletes everything ready to start again.

$\boxed{CE}$ for Cancel Entry or $\boxed{DEL}$ for DELete, both delete the last key that was pressed.

$\boxed{ANS}$ enters the Answer that is on the display, ready for another calculation. For example:

$\boxed{2}$ $\boxed{\times}$ $\boxed{3}$ $\boxed{=}$ gives a display of 6.

$\boxed{ANS}$ $\boxed{\times}$ $\boxed{4}$ $\boxed{=}$ gives 24, because it works out 6×24.

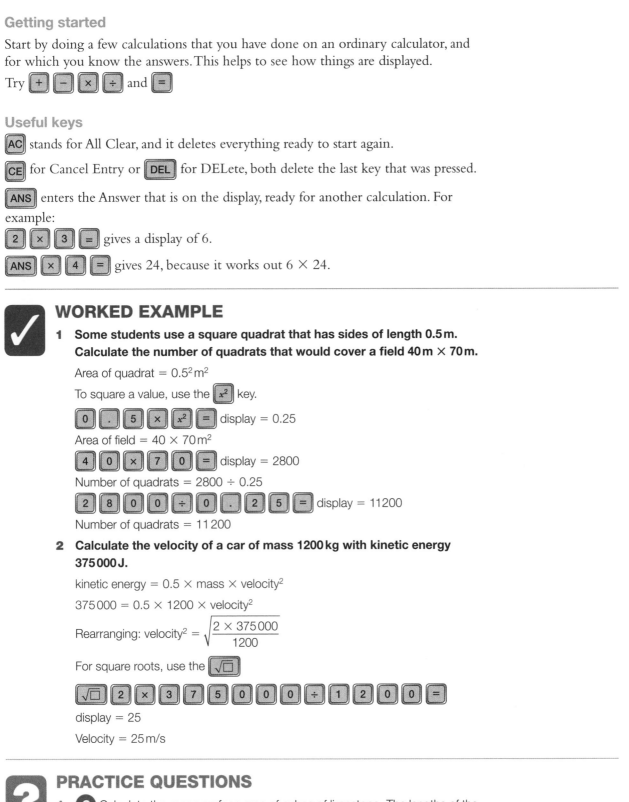

✔ WORKED EXAMPLE

1 Some students use a square quadrat that has sides of length 0.5 m. Calculate the number of quadrats that would cover a field 40 m × 70 m.

Area of quadrat $= 0.5^2 \, \text{m}^2$

To square a value, use the $\boxed{x^2}$ key.

$\boxed{0}$ $\boxed{.}$ $\boxed{5}$ $\boxed{\times}$ $\boxed{x^2}$ $\boxed{=}$ display = 0.25

Area of field $= 40 \times 70 \, \text{m}^2$

$\boxed{4}$ $\boxed{0}$ $\boxed{\times}$ $\boxed{7}$ $\boxed{0}$ $\boxed{=}$ display = 2800

Number of quadrats $= 2800 \div 0.25$

$\boxed{2}$ $\boxed{8}$ $\boxed{0}$ $\boxed{0}$ $\boxed{\div}$ $\boxed{0}$ $\boxed{.}$ $\boxed{2}$ $\boxed{5}$ $\boxed{=}$ display = 11200

Number of quadrats $= 11\,200$

2 Calculate the velocity of a car of mass 1200 kg with kinetic energy 375 000 J.

kinetic energy $= 0.5 \times \text{mass} \times \text{velocity}^2$

$375\,000 = 0.5 \times 1200 \times \text{velocity}^2$

Rearranging: $\text{velocity}^2 = \sqrt{\dfrac{2 \times 375\,000}{1200}}$

For square roots, use the $\boxed{\sqrt{\square}}$

$\boxed{\sqrt{\square}}$ $\boxed{2}$ $\boxed{\times}$ $\boxed{3}$ $\boxed{7}$ $\boxed{5}$ $\boxed{0}$ $\boxed{0}$ $\boxed{0}$ $\boxed{\div}$ $\boxed{1}$ $\boxed{2}$ $\boxed{0}$ $\boxed{0}$ $\boxed{=}$

display = 25

Velocity = 25 m/s

? PRACTICE QUESTIONS

1 **C** Calculate the mean surface area of cubes of limestone. The lengths of the sides are: 0.5 cm, 0.6 cm, 0.5 cm, 0.6 cm, 0.4 cm, 0.5 cm, 0.4 cm.

2 **B** Calculate the surface area of a cube with sides 4.5 cm.

6.2 Using a calculator 2

More useful functions

If you have a key $\boxed{x^3}$ then this can be used to cube values.

To calculate $\dfrac{1}{x}$ use $\boxed{x^{-1}}$ because $x^{-1} = \dfrac{1}{x}$

If you miss out a character, you can use $\boxed{\blacktriangleleft}$ to go back through the keys you have entered. When you have pressed it enough times to get to the right point, press the key you want to enter the missing character.

$\boxed{\blacktriangleright}$ will go forward through the characters on the display, but you can just key $\boxed{=}$ without going to the end of the line.

✓ WORKED EXAMPLES

1 Calculate the volume of a cube with sides of 14 cm.

Volume = 14^3

$\boxed{1}\boxed{4}\boxed{x^3}\boxed{=}$ display = 2744 cm³

2 You have been asked to plot a graph of light intensity against $\dfrac{1}{distance^2}$, where distance = the distance of the lamp source to a plant.

The first value of distance = 25 cm.

Key in $\boxed{2}\boxed{5}\boxed{x^2}\boxed{=}\boxed{\text{ANS}}\boxed{x^{-1}}\boxed{=}$ display = 1.6×10^{-3}/cm²

Notice that you have to key $\boxed{=}\boxed{\text{ANS}}$ because the calculator does not allow you to do $\boxed{x^2}\boxed{x^{-1}}$. See Topic 6.1 for more about using $\boxed{\text{ANS}}$.

? PRACTICE QUESTIONS

1 (P) The mass of a cube of aluminium with sides of 0.3 m is 72.9 kg. Calculate **a** the volume of the cube and **b** the density of aluminium in kg/m³.

2 (C) Calculate the surface area to volume ratio of cubes of limestone with length of sides a 0.8 cm and b 0.4 cm.

3 (P) You have been asked to plot a graph of pressure against $\dfrac{1}{volume}$ for a gas.

Calculate the values you need to plot. Complete the table below.

volume in cm³	10	15	20	25
$\dfrac{1}{volume}$ in 1/cm³				

4 (B) Complete the table for a graph of light intensity against $\dfrac{1}{distance^2}$ by calculating the values.

distance in cm	12	24	36	48
$\dfrac{1}{distance^2}$ in 1/cm²				

BODMAS

BODMAS stands for the order in which the steps in a calculation must be done.

Step 1: Brackets – work out the sums inside brackets first.

Step 2: Orders – this means powers, such as squares, square roots and cubes.

Step 3: Divide

Step 4: Multiply

Step 5: Add

Step 6: Subtract

WORKED EXAMPLE

Determine the slope of a line between two points on a graph with co-ordinates (2,6) and (4,7).

You need to work out: slope $= \dfrac{7 - 6}{4 - 2}$

The line under $7 - 6$ acts like brackets around $(7 - 6)$ and also around $(4 - 2)$, so these two subtraction sums must be done before the division.

Writing $\dfrac{(7 - 6)}{(4 - 2)}$ makes this clearer.

Key in: (7 − 6) ÷ (4 − 2) =

answer on display $= 0.5$

Note that if you key in: 7 − 6 ÷ 4 − 2 = this is $7 - \dfrac{6}{4} - 2 = 3.5$, which is incorrect.

Another way to do this calculation is to enter it as a fraction (see Topic 6.3).

PRACTICE QUESTIONS

5 **P** Calculate the velocity of a car with kinetic energy 285 560 J. The car has mass 1000 kg and the passengers and driver together have mass 180 kg.

6 **C** A student plotted a graph of concentration (mol/dm^3) against time (s). Calculate the gradient of the tangent to the curve 20 s after the reaction starts. The tangent passes through the points (0,0.04) and (46,0.5).

   a Calculate the slope of the tangent.

   b State the rate of reaction.

7 **B** A student's mass when wearing clothes is 61 kg. The clothes have a mass of 2 kg. The student's BMI is 21.7.

   Use the equation $BMI = \dfrac{mass}{height^2}$ to calculate their height.

6.3 Using a calculator 3

Standard form

To enter a number in standard form, use the $\boxed{\times 10^x}$ key.

For example, the Avogadro constant $= 6.02 \times 10^{23}$ per mole.

To enter this, key in $\boxed{6}\ \boxed{.}\ \boxed{0}\ \boxed{2}\ \boxed{\times 10^x}\ \boxed{2}\ \boxed{3}$

This can be very useful.

You can also use it with prefixes like milli and mega to convert the values for calculations.

> **!** REMEMBER: You do not need to use the $\boxed{\times}\ \boxed{1}\ \boxed{0}$ keys to enter a number in standard form. Instead, use $\boxed{\times 10^x}$.

WORKED EXAMPLES

1 **The wavelength of blue light is 4.5×10^{-7} m and the speed of light $= 3.0 \times 10^8$ m/s. Calculate the frequency. State your answer in standard form.**

Velocity = frequency × wavelength

$3.0 \times 10^8 = 4.5 \times 10^{-7} \times$ frequency

Frequency $= \dfrac{3.0 \times 10^8}{4.5 \times 10^{-7}}$

Key in $\boxed{3}\ \boxed{\times 10^x}\ \boxed{8}\ \boxed{\div}\ \boxed{4}\ \boxed{.}\ \boxed{5}\ \boxed{\times 10^x}\ \boxed{-}\ \boxed{7}\ \boxed{=}$

display $= 6.66667 \times 10^{14}$

Notice that you don't need to enter $\boxed{.}\ \boxed{0}$ – it won't make any difference to the calculation.

Answer $= 6.7 \times 10^{14}$ m/s

2 **A virus has a length 110 nm. It is viewed with a magnification × 10 000. Calculate the length of the image. State your answer in standard form.**

This is an example of how a scientific calculator can be very useful.

You need to know that nano $= 10^{-9}$

Key in $\boxed{1}\ \boxed{1}\ \boxed{0}\ \boxed{\times 10^x}\ \boxed{-}\ \boxed{9}\ \boxed{\times}\ \boxed{1}\ \boxed{0}\ \boxed{0}\ \boxed{0}\ \boxed{0}\ \boxed{=}$

display $= 1.1 \times 10^{-3}$

PRACTICE QUESTIONS

1 **C** A nanoparticle is a cube with sides of 0.7 nm. Calculate **a** the surface area of the particle and **b** the volume. Give your answers in standard form.

2 **P** A radio wave has frequency 98 MHz. Use the fact that radio waves travel at 3×10^8 m/s to calculate the wavelength.

3 **B** There are an estimated 37.2×10^{12} cells in the human body. About 80% of these are red blood cells. Calculate an estimate of the number of red blood cells. Give your answer in standard form to 2 s.f.

Using fractions

You can enter the calculation as it looks on the page as a fraction. The key for this is 🔳.

Whether you enter numbers as fractions or decimals is up to you, but your answers must be in decimals.

There will usually be a final mark for the correct answer. You will not gain this mark if the answer is not calculated but is left as a fraction.

You can change an answer from a fraction to a decimal by using the key $\boxed{S{\Leftrightarrow}D}$

This also changes the answer from a decimal to a fraction.

✓ WORKED EXAMPLES

3 **Calculate** $\dfrac{1}{2}+\dfrac{3}{4}$. **Give your answer as a decimal.**

Key in $\boxed{1}$ $\boxed{\blacksquare}$ $\boxed{2}$ $\boxed{\blacktriangleright}$ $\boxed{+}$ $\boxed{3}$ $\boxed{\blacksquare}$ $\boxed{4}$ $\boxed{=}$

The answer displayed will be $\dfrac{5}{4}$ or 1.25

If it is $\dfrac{5}{4}$ key in $\boxed{S{\Leftrightarrow}D}$. The answer displayed will be 1.25

4 **The speed of a sound wave = 330 m/s and the frequency = 360 Hz.**
Calculate the wavelength.

Velocity = frequency × wavelength

Rearranging: $\dfrac{\text{velocity}}{\text{frequency}} = \dfrac{330}{360}$

Key in $\boxed{3}$ $\boxed{3}$ $\boxed{0}$ $\boxed{\blacksquare}$ $\boxed{3}$ $\boxed{6}$ $\boxed{0}$ $\boxed{=}$. The answer displayed $= \dfrac{11}{12}$

Key in $\boxed{S{\Leftrightarrow}D}$. The answer displayed = 0.916. Wavelength = 0.916 m

The answer displayed has a dot over the final 6. This means the answer is a recurring decimal – and if you press $\boxed{S{\Leftrightarrow}D}$ again, the answer is displayed rounded to a number of decimal places. Pressing $\boxed{S{\Leftrightarrow}D}$ a third time goes back to a fraction.

? PRACTICE QUESTIONS

4 Ⓒ Calculate the concentration of a solution made with 3.5 g of sodium chloride dissolved in 0.45 dm³ of water.

5 Ⓑ A cell 25 μm in diameter is magnified and the image diameter = 10 mm. Calculate the magnification.

6 Ⓟ Calculate the power of a crane motor that lifts a weight of 260 000 N up 25 m in 48 s.

Use the equations:

$\text{power} = \dfrac{\text{work done}}{\text{time taken}}$ and $\text{work done} = \text{force} \times \text{distance}$

Lists of equations

Equation	In symbols	Exam board Edexcel = E OCR = O AQA = A
Biology		
magnification $= \dfrac{\text{image size}}{\text{object size}}$	$M = \dfrac{I}{O}$	E O A
$BMI = \dfrac{\text{mass in kg}}{(\text{height in m})^2}$		E
Fick's law of diffusion: rate of diffusion $\propto \dfrac{\text{surface area} \times \text{concentration difference}}{\text{thickness of membrane}}$		E
cardiac output $=$ stroke volume $\times$ heart rate		E
light intensity $\propto \dfrac{1}{\text{distance from point light source}^2}$		E A O
Chemistry		
relative formula mass of molecule $=$ sum of relative atomic masses of the atoms		A E O
% mass of element in compound $= \dfrac{\text{relative mass of element}}{\text{relative formula mass of compound}} \times 100\%$		A E O
relative atomic mass $=$ sum of $(\dfrac{\text{\% abundance}}{100} \times$ relative atomic mass of isotope) for all isotopes		A E O
number of moles $= \dfrac{\text{mass (g)}}{\text{relative formula mass}}$		A E O
% yield $= \dfrac{\text{actual yield}}{\text{maximum theoretical yield}} \times 100\%$		A E O
% atom economy $= \dfrac{\text{relative formula mass of desired product from equation}}{\text{sum of relative formula masses of all reactants from equation}} \times 100\%$		A E O
number of molecules $=$ number of moles $\times\ 6.02 \times 10^{23}$ per mole		A E O
concentration (g/dm^3) $= \dfrac{\text{solute (g)}}{\text{solvent (dm}^3)}$		A E O
concentration (mol/dm^3) $= \dfrac{\text{solute (mol)}}{\text{solvent (dm}^3)}$		A E O
gas volume $=$ number of moles $\times$ 24 dm^3 (at room temperature and pressure)		A E O
mean rate of reaction $= \dfrac{\text{quantity of reactant used}}{\text{time}}$		A E O
mean rate of reaction $= \dfrac{\text{quantity of product formed}}{\text{time}}$		A E O
$R_f = \dfrac{\text{distance moved by substance}}{\text{distance moved by solvent}}$		A E O

Physics

Description	Equation	
velocity $= \dfrac{\text{displacement}}{\text{time}}$ and speed $= \dfrac{\text{distance}}{\text{time}}$	$v = \dfrac{s}{t}$	A E O
acceleration $= \dfrac{\text{change in velocity}}{\text{time}}$	$a = \dfrac{(v - u)}{t}$	A E O
final velocity squared $-$ initial velocity squared $= 2 \times$ acceleration $\times$ distance	$v^2 - u^2 = 2as$	A E O
resultant force $=$ mass $\times$ acceleration	$F = ma$	A E O
weight $=$ mass $\times$ gravitational field strength	$W = mg$	A E O
work done $=$ force $\times$ distance	$W = Fd$	A E O
kinetic energy $= 0.5 \times$ mass $\times$ (speed)^2	$KE = \dfrac{1}{2}mv^2$	A E O
potential energy $=$ mass $\times$ gravitational field strength $\times$ change in height	$PE = mgh$	A E O
power $= \dfrac{\text{work done}}{\text{time taken}} = \dfrac{\text{energy transferred}}{\text{time taken}}$	$P = \dfrac{W}{t} = \dfrac{E}{t}$	A E O
efficiency $= \dfrac{\text{useful power output}}{\text{total power input}} \times 100\%$ or $\dfrac{\text{useful energy output}}{\text{total energy input}} \times 100\%$		A E O
momentum $=$ mass $\times$ velocity	$p = mv$	A E O
force $= \dfrac{\text{change in momentum}}{\text{time taken}}$	$F = \dfrac{(mv - mu)}{t}$	A E O
density $= \dfrac{\text{mass}}{\text{volume}}$	$\rho = \dfrac{m}{V}$	A E O
pressure $= \dfrac{\text{force}}{\text{area}}$	$P = \dfrac{F}{A}$	A E O
Hooke's law: force $\propto$ extension force $=$ spring constant $\times$ extension	$F = kx$	A E O
elastic potential energy $= 0.5 \times$ spring constant $\times$ (extension)^2	$E_e = \dfrac{1}{2}Kx^2$	A E O
resistance $= \dfrac{\text{potential difference}}{\text{current}}$	$R = \dfrac{V}{I}$	A E O
total resistance in series $=$ sum of resistances	$R_T = R_1 + R_2 + R_3$	A
total resistance in parallel: $\dfrac{1}{\text{total resistance}} = \dfrac{1}{\text{resistance 1}} = \dfrac{1}{\text{resistance 2}} = \dfrac{1}{\text{resistance 3}}$	$\dfrac{1}{R_T} = \dfrac{1}{R_1} + \dfrac{1}{R_2} + \dfrac{1}{R_3}$	A
charge flow $=$ current $\times$ time	$Q = It$	A E O
energy transferred $=$ charge flow $\times$ potential difference	$E = QV$	A E O
power $=$ potential difference $\times$ current	$P = IV$	A E O
power $=$ (current)$^2 \times$ resistance	$P = I^2R$	A E O
energy transferred $=$ power $\times$ time	$E = Pt$	A E O
magnetic force on a wire $=$ magnetic flux density $\times$ current $\times$ length of wire	$F = BIl$	A E O
$\dfrac{\text{potential difference across primary coil}}{\text{potential difference across secondary coil}} = \dfrac{\text{number of turns in primary coil}}{\text{number of turns in secondary coil}}$	$\dfrac{V_P}{V_s} = \dfrac{n_p}{n_s}$	A E O

pd across primary coil × current in primary coil = pd across secondary coil × current in secondary coil	$V_p I_p = V_s I_s$	A E O
wave speed = frequency × wavelength	$V = f\lambda$	A E O
frequency = $\dfrac{1}{\text{period}}$	$F = \dfrac{1}{T}$	A E O
For a gas: pressure is inversely proportional to volume *or* pressure × volume = constant	$p \propto \dfrac{1}{V}$ $p_1 V_1 = p_2 V_2$	A E O
change in thermal energy = mass × specific heat capacity × temperature change	$E = m C (\theta_2 - \theta_1)$	A E O
thermal energy for a change of state = mass × specific latent heat	$E = m L$	A E O
magnification = $\dfrac{\text{image height}}{\text{object height}}$	$M = \dfrac{I}{O}$	A O
pressure due to a column of liquid = height of column × density of liquid × gravitational field strength	$p = h \rho g$	A E O
moment of a force = force × distance normal to the direction of the force		E O

Maths

circumference of a circle = $2 \times \pi \times$ radius

area of a square = (length of side)^2

area of a rectangle = length × width

area of a triangle = $0.5 \times$ base × perpendicular height

area of a circle = π (radius)^2

cross-sectional area = area of slice through object

surface area = total area of faces

surface area of cube = $6 \times$ length^2

volume of a cube = length^3

volume of a cuboid = length × width × height

Periodic table

key

| relative atomic mass |
| **atomic symbol** |
| name |
| atomic (proton) number |

1	1
H	
hydrogen	
1	

1	2												3	4	5	6	7	0
																		4 **He** helium 2
7 **Li** lithium 3	9 **Be** beryllium 4												11 **B** boron 5	12 **C** carbon 6	14 **N** nitrogen 7	16 **O** oxygen 8	19 **F** fluorine 9	20 **Ne** neon 10
23 **Na** sodium 11	24 **Mg** magnesium 12												27 **Al** aluminium 13	28 **Si** silicon 14	31 **P** phosphorus 15	32 **S** sulfur 16	35.5 **Cl** chlorine 17	40 **Ar** argon 18
39 **K** potassium 19	40 **Ca** calcium 20	45 **Sc** scandium 21	48 **Ti** titanium 22	51 **V** vanadium 23	52 **Cr** chromium 24	55 **Mn** manganese 25	56 **Fe** iron 26	59 **Co** cobalt 27	59 **Ni** nickel 28	63.5 **Cu** copper 29	65 **Zn** zinc 30	70 **Ga** gallium 31	73 **Ge** germanium 32	75 **As** arsenic 33	79 **Se** selenium 34	80 **Br** bromine 35	84 **Kr** krypton 36
85 **Rb** rubidium 37	88 **Sr** strontium 38	89 **Y** yttrium 39	91 **Zr** zirconium 40	93 **Nb** niobium 41	96 **Mo** molybdenum 42	[98] **Tc** technetium 43	101 **Ru** ruthenium 44	103 **Rh** rhodium 45	106 **Pd** palladium 46	108 **Ag** silver 47	112 **Cd** cadmium 48	115 **In** indium 49	119 **Sn** tin 50	122 **Sb** antimony 51	128 **Te** tellurium 52	127 **I** iodine 53	131 **Xe** xenon 54
133 **Cs** caesium 55	137 **Ba** barium 56	139 **La*** lanthanum 57	178 **Hf** hafnium 72	181 **Ta** tantalum 73	184 **W** tungsten 74	186 **Re** rhenium 75	190 **Os** osmium 76	192 **Ir** iridium 77	195 **Pt** platinum 78	197 **Au** gold 79	201 **Hg** mercury 80	204 **Tl** thallium 81	207 **Pb** lead 82	209 **Bi** bismuth 83	[209] **Po** polonium 84	[210] **At** astatine 85	[222] **Rn** radon 86
[223] **Fr** francium 87	[226] **Ra** radium 88	[227] **Ac*** actinium 89	[261] **Rf** rutherfordium 104	[262] **Db** dubnium 105	[266] **Sg** seaborgium 106	[264] **Bh** bohrium 107	[277] **Hs** hassium 108	[268] **Mt** meitnerium 109	[271] **Ds** darmstadtium 110	[272] **Rg** roentgenium 111	[285] **Cn** copernicium 112	[286] **Nh** nihonium 113	[289] **Fl** flerovium 114	[289] **Mc** moscovium 115	[293] **Lv** livermorium 116	[294] **Ts** tennessine 117	[294] **Og** oganesson 118

*The lanthanides (atomic numbers 58–71) and the actinides (atomic numbers 90–103) have been omitted.

Relative atomic masses for **Cu** and **Cl** have not been rounded to the nearest whole number.

Index

Notes

Notes

Notes

Notes